Springer Theses

Recognizing Outstanding Ph.D. Research

Aims and Scope

The series "Springer Theses" brings together a selection of the very best Ph.D. theses from around the world and across the physical sciences. Nominated and endorsed by two recognized specialists, each published volume has been selected for its scientific excellence and the high impact of its contents for the pertinent field of research. For greater accessibility to non-specialists, the published versions include an extended introduction, as well as a foreword by the student's supervisor explaining the special relevance of the work for the field. As a whole, the series will provide a valuable resource both for newcomers to the research fields described, and for other scientists seeking detailed background information on special questions. Finally, it provides an accredited documentation of the valuable contributions made by today's younger generation of scientists.

Theses may be nominated for publication in this series by heads of department at internationally leading universities or institutes and should fulfill all of the following criteria

- They must be written in good English.
- The topic should fall within the confines of Chemistry, Physics, Earth Sciences, Engineering and related interdisciplinary fields such as Materials, Nanoscience, Chemical Engineering, Complex Systems and Biophysics.
- The work reported in the thesis must represent a significant scientific advance.
- If the thesis includes previously published material, permission to reproduce this must be gained from the respective copyright holder (a maximum 30% of the thesis should be a verbatim reproduction from the author's previous publications).
- They must have been examined and passed during the 12 months prior to nomination.
- Each thesis should include a foreword by the supervisor outlining the significance of its content.
- The theses should have a clearly defined structure including an introduction accessible to new PhD students and scientists not expert in the relevant field.

Indexed by zbMATH.

More information about this series at https://link.springer.com/bookseries/8790

Ayumi Imayoshi

Discrimination of Mobile Supramolecular Chirality

Acylative Molecular Transformations by Organocatalysis

Doctoral Thesis accepted by Kyoto University, Kyoto, Japan

Springer

Author
Dr. Ayumi Imayoshi
Kyoto University
Kyoto, Japan

Supervisor
Prof. Takeo Kawabata
Institute for Chemical Research
Kyoto University
Kyoto, Japan

ISSN 2190-5053 ISSN 2190-5061 (electronic)
Springer Theses
ISBN 978-981-16-7430-3 ISBN 978-981-16-7431-0 (eBook)
https://doi.org/10.1007/978-981-16-7431-0

This Springer imprint is published by the registered company Springer Nature Singapore Pte Ltd.
The registered company address is: 152 Beach Road, #21-01/04 Gateway East, Singapore 189721, Singapore

Supervisor's Foreword

It is my great pressure to introduce Dr. Ayumi Imayoshi on the occasion of Springer Thesis Prize as an outstanding original doctoral work. Dr. Imayoshi joined our group at the Institute for Chemical Research, Kyoto University, in April 2010 as an undergraduate student. After graduation from Kyoto University, she entered the Graduate School of Pharmaceutical Sciences of Kyoto University, started her doctoral study with me and received her Ph.D. degree from Kyoto University in 2016.

Dr. Imayoshi's research concerns asymmetric synthesis of chiral mechanically interlocked molecules. Asymmetric synthesis has been extensively developed, and the field of asymmetric synthesis is becoming a mature science. On the other hand, asymmetric synthesis of chiral mechanically interlocked molecules such as mechanically planar chiral rotaxanes and topologically chiral catenanes has scarcely been achieved. Since the discovery of such supramolecular chirality nearly 50 years ago, its enantioselective construction is a long-standing dream in organic synthesis. The pioneering first report on enantioselective synthesis of a mechanically planar chiral rotaxane was reported by Takata in 2007. However, the reported maximum enantioselectivity was only 4.4% ee. This has been the only example for enantioselective construction of mechanically planar chiral rotaxanes or topologically chiral catenanes as of 2015 when she finished her Ph.D. study. Why asymmetric synthesis of such molecules has been so difficult? We believe that the diversity and mobility of chiral mechanically interlocked molecules could be the reasons.

In December 2012, Dr. Imayoshi and I discussed to overcome this problem and reached the conclusion that efficient kinetic resolution of racemic mechanically planar chiral rotaxanes would be achievable by employing our original catalysts with excellent ability for recognition of remote chirality. Dr. Imayoshi devoted tremendous efforts to pursue this objective. She developed her original protocol for kinetic resolution of racemic mechanically planar chiral rotaxanes by asymmetric acylation to finally provide the rotaxanes in >99% ee at 71 conversion (selectivity factor > 14.5). This is the first example of highly enantioselective construction of chiral mechanically interlocked molecules, and the results were published in Nature Communications as she is the first author (A. Imayoshi *et al.*, *Nat. Commun.* **2021**, *12*, 404). The salient feature is that efficient discrimination of mechanically planar chirality of the substrate rotaxane was achieved by unusually effective remote asymmetric acylation of the hydroxy group located at the edge of the huge rotaxane molecule via precise molecular recognition with our catalyst. It is also worthy to note that a most simple organic reaction, acetylation of alcohol, solved the most challenging task in organic synthesis.

The success of the unprecedented work is obviously relied on Dr. Imayoshi's continuous efforts and her intelligence. I respect her as a researcher as well as a person. She has never forgotten her smile all the time even when the study was facing to very hard situations. I congratulate her great challenging mind to realize a dream of organic chemistry and sincerely wish her further success in research.

Kyoto Japan
March 2021

Takeo Kawabata

List of Published Articles

Parts of this thesis have been published in the following journal articles:

1. (Chapter 3)
   Enantioselective preparation of mechanically planar chiral rotaxanes by kinetic resolution strategy
   Ayumi Imayoshi, Bhatraju Vasantha Lakshmi, Yoshihiro Ueda, Tomoyuki Yoshimura, Aki Matayoshi, Takumi Furuta, Takeo Kawabata
   Nature Communications, **12**, 404 (2021).

2. (Chapter 2)
   Insights into the Molecular Recognition Process in Organocatalytic Chemoselective Monoacylation of 1,5-Pentanediol
   Ayumi Imayoshi, Masahiro Yamanaka, Makoto Sato, Keisuke Yoshida, Takumi Furuta, Yoshihiro Ueda, Takeo Kawabata
   Advanced Synthesis & Catalysis, **358**, 1337–1344 (2016).

3. (Chapter 4)
   Carboxylate Anions Accelerate Pyrrolidinopyridine (PPy)-Catalyzed Acylation: Catalytic Site-Selective Acylation of a Carbohydrate by in Situ Counteranion Exchange
   Masanori Yanagi, Ayumi Imayoshi, Yoshihiro Ueda, Takumi Furuta, Takeo Kawabata
   Organic Letters, **19**, 3099–3102 (2017).

Acknowledgements

Firstly, I express my sincere gratitude to Prof. Takeo Kawabata (Kyoto University) for his generous discussion, thoughtful guidance, support and help throughout the research. I am truly happy to have been able to lay the foundation as a researcher, learning his attractive research perspective and personality, over six years on the first stage of my research life.

I am thankful to Prof. Takumi Furuta (Kyoto Pharmaceutical University) for his valuable and useful discussions and suggestions. I wish to thank Associate Professor Tomoyuki Yoshimura (Kanazawa University) for discussing my ideas and sharing his sophisticated skills, knowledge and experience. I am grateful to Assistant Professor Yoshihiro Ueda (Kyoto University) for useful discussion on a wide range of matters.

I am really grateful to Prof. Masahiro Yamanaka (Rikkyo University) and Dr. Makoto Sato for their substantial contributions to DFT calculations and useful discussions. I would like to thank Associate Professor Kazuto Takaishi (Okayama University) and Professor Kenji Monde (Hokkaido University) for calculations of supramolecules. I wish to thank Prof. Kazunori Tsubaki (Kyoto Prefectural University) for helpful advice and giving me a valuable research opportunity.

I would like to thank Prof. Shiro Futaki (Kyoto University) and Prof. Ikuhiko Nakase (Osaka Prefecture University) for their kind guidance in measurement of MALDI mass spectrometry, Prof. Masaharu Nakamura (Kyoto University) and Associate Professor Hikaru Takaya (Kyoto University) for their helpful guidance in measurement of ESI mass spectrometry, Ms. Kyoko Ohmine for measuring variable-temperature NMR spectra and Ms. Akiko Fujihashi for mass spectrometry.

I would like to express my appreciation to Prof. Kiyosei Takasu (Kyoto University) and Prof. Yoshiji Takemoto (Kyoto University) for reviewing my doctoral thesis and for providing valuable comments.

I wish to thank Prof. Jun Hiratake (Kyoto University) for his useful guidance in research ethics. I gratefully thank all members of his seminar for enjoyable discussion.

I would like to thank Prof. Ivan Huc (University of Bordeaux) and all members of Huc group for giving me the valuable opportunity to learn their excellent chemistry. My stay in France was supported by the Kyoto University Foundation.

I am grateful to Dr. Bhatraju Vasantha Lakshmi, Dr. Keisuke Yoshida, Dr. Keisuke Tomohara, Dr. Kenji Mishiro, Dr. Shohei Hamada, Dr. Takashi Shigeta and Dr. Masanori Yanagi for their valuable discussions and suggestions. I express my gratitude to Dr. Hironori Takeuchi, Dr. Ryuichi Hyakutake and Dr. Atsushi Hirata for their kind help and constant encouragement over several years. I am thankful to the secretary, Ms. Kaori Hashimoto, for her support. I gratefully thank all past and present members of Kawabata Group for active discussions and sharing various experiences.

This study was supported by Japan Society for the Promotion of Science (JP15J10954).

I express my deep gratitude to my parents for their kind support, understanding and continuous encouragement.

Finally, I thank my husband, Ryuji, and my two daughters, Misaki and Emiri, with all my heart. They have always made me happy, and words cannot express how grateful I am. I would like to share many adventures and dreams together in the future too.

Contents

Abbreviations

1D	One dimension
2D	Two dimensions
Ac	Acetyl
aq.	Aqueous
Boc	*Tert*-butyloxycarbonyl
Bu	Butyl
Bz	Benzoyl
C	Concentration
C	Conversion
cat.	Catalyst
CD	Circular dichroism
COSY	Correlation spectroscopy
dba	Dibenzylideneacetone
DCC	*N,N′*-dicyclohexylcarbodiimide
de	Diastereomeric excess
DEPT	Distortionless enhancement by polarization transfer
DFT	Density functional theory
DMAP	4-dimethylaminopyridine
DMF	*N,N*-dimethylformamide
DMSO	Dimethylsulfoxide
dppp	1,3-bis(diphenylphosphino)propane
e.g.	For example
ee	Enantiomeric excess
ent	Enantio-
eq	Equivalent
Et	Ethyl
G	Gibbs free energy
H	Enthalpy
h	Hour
HMBC	Heteronuclear multiple bond coherence
HMQC	Heteronuclear multiple quantum correlation

HPLC	High-performance liquid chromatography
HRMS	High-resolution mass spectrometer
IBX	2-iodoxybenzoic acid
^{i}Pr	Isopropyl
IR	Infrared
K	Equilibrium constant
K	Kelvin
k	Rate constant
m.p.	Melting point
Me	Methyl
min	Minute
NMR	Nuclear magnetic resonance
Ns	2-nitrobenzenesulfonyl
o-	Ortho-
p-	Para-
PPTS	Pyridinium *p*-toluenesulfonate
PPY	4-pyrrolidinopyridine
PTLC	Preparative thin-layer chromatography
rt	Room temperature
s factor	Selectivity factor
S	Entropy
sat.	Saturated
SM	Starting material
^{t}Bu	*Tert*-butyl
TD-DFT	Time-dependent density functional theory
temp.	Temperature
tert	Tertiary
Tf	Trifluoromethanesulfonyl
THF	Tetrahydrofuran
THP	Tetrahydropyran
TLC	Thin-layer chromatography
Ts	*p*-toluenesulfonyl
TS	Transition state
UV	Ultraviolet
XPhos	2-dicyclohexylphosphino-2′,4′,6′-triisopropylbiphenyl
ZPE	Zero-point energy

Chapter 1
Introduction

Abstract Supramolecules such as rotaxanes and catenanes are known to possess unique supramolecular chirality (Fig. 1.1). Mechanically planar chirality and topological chirality in the structure of rotaxanes and catenanes have mobility in the chiral environment themselves. The chiral environment is kinetically labile since the two components in the structure of interlocked molecules can move freely from each other. In this book, the mobile nature of the chiralities of rotaxanes and catenanes is referred to as "*mobile supramolecular chirality*". I will discuss how to discriminate the mobile supramolecular chirality by catalysts and achieve the production of enantiopure mechanically planar chiral rotaxanes.

Keywords Mobile supramolecular chirality · Rotaxanes and catenanes · Molecular transformations · Catalyst

In the field of organic chemistry, asymmetric synthesis and chiral molecular transformation have been extensively developed during the last few decades. Beginning with the discovery of a proline catalyst [1–6], asymmetric synthesis has grown into one of the main research areas of modern organic chemistry. Catalytic asymmetric transformations have contributed to wide range of fields including medicinal chemistry and biochemistry because majority of organic compounds associated with living organisms are chiral.

Various kinds of chiralities are shown in Fig. 1.2 [7–18]. Asymmetric construction of conventional chiralities such as central chirality, axial chirality [19–23], planar chirality [19], and helical chirality [24–28] have already been actively studied. With the current state of chemical science, asymmetric synthesis of these fixed chiralities over covalent bonds is reaching a level of mature science.

On these fields approaching mature science, however, asymmetric synthesis of supramolecules still stands as an unsolved problem. Supramolecules such as rotaxanes and catenanes, consisting of more than two components linked non-covalent system [29], are known to possess unique supramolecular chirality (Fig. 1.1) [30–32]. Even though each component has no chirality, their combination brings mechanically linked chirality, when each of the component has structural dissymmetry. Such chirality generated from the orientation of the components is called mechanical chirality.

A. Imayoshi, *Discrimination of Mobile Supramolecular Chirality*, Springer Theses,
https://doi.org/10.1007/978-981-16-7431-0_1

Fig. 1.1 Mobile supramolecular chirality generated from the orientation of two dissymmetric components linked non-covalent system

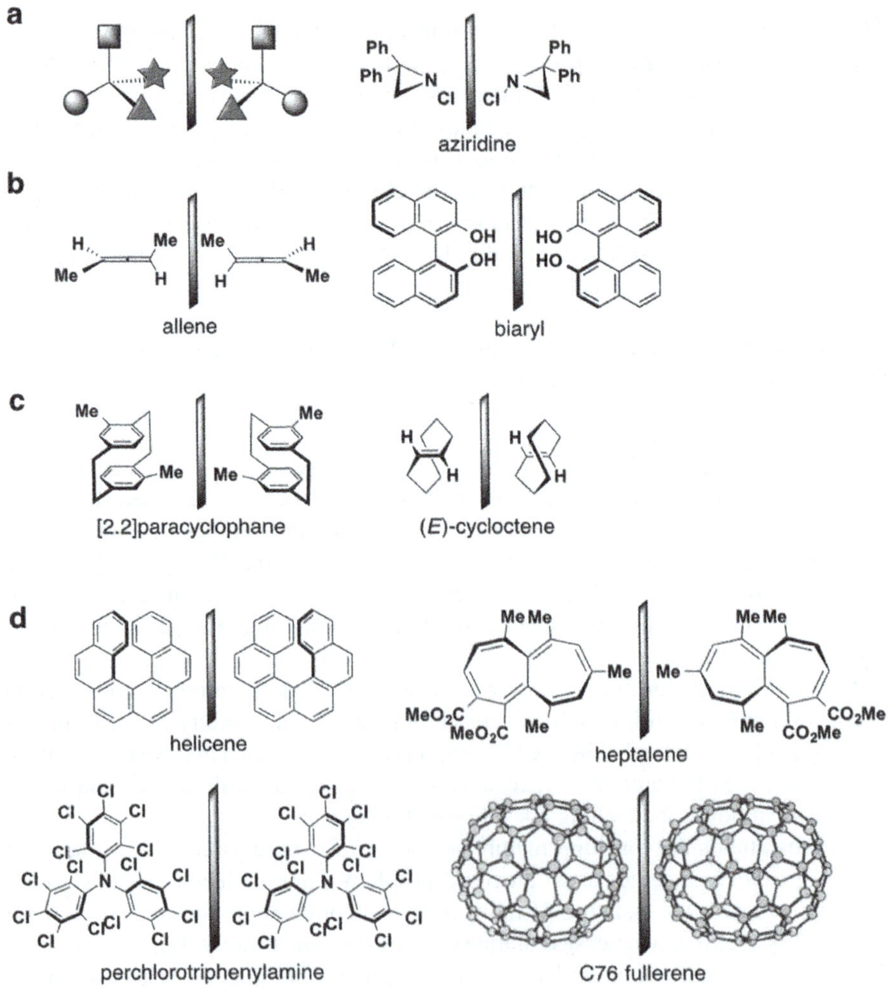

Fig. 1.2 Various kinds of chiralities, conventional fixed chiralities over covalent bonds. **a** Central chirality, **b** axial chirality, **c** planar chirality, **d** helical chirality

The difference between mechanical chirality of supramolecules and other chiralities is the mobility of the mechanical chirality itself. Since two components can move freely from each other, the mechanical chirality, generated from the combination of these two phases, has mobility in the chiral environment itself. A molecule possessing mechanical chirality continuously changes its chiral environment with kinetic liability. Thus, mechanical chirality of rotaxanes and catenanes can be called *mobile supramolecular chirality*.

This mobile nature of mechanical chirality makes asymmetric synthesis of mechanically chiral molecules much more difficult than the compounds with conventional chiralities. In fact, traditional molecular transformations do not work efficiently for the construction of the molecules with mobile supramolecular system. The most successful and pioneering example of catalytic asymmetric synthesis of mechanically chiral supramoelcules so far achieved gave the rotaxane in up to only 4.4% ee [33]. Thus, novel concept and method for the molecular transformation are required in organic chemistry for achieving asymmetric synthesis of supramolecules.

Toward discrimination of mobile supramolecular chirality, I first investigated the fundamental nature of catalytic acylation. In our laboratory (Kawabata group), various acylative molecular transformations have been developed by using chiral 4-pyrrolidinopyridine (PPY) catalysts which bear substrate-recognition sites consisting of amino acid side chains [34–39]. Among them, I focused on the most simple and fundamental molecular transformation [36], and elucidated the molecular recognition process promoted by the chiral PPY catalyst (Chap. 2).

Then, I challenged asymmetric synthesis of supramolecules by using the chiral PPY catalyst. I achieved the first example of highly selective asymmetric discrimination of mobile supramolecular chirality (Chap. 3).

In this book, I propose the concept of *dynamic molecular recognition* by catalysts for discrimination of mobile supramolecular chirality. This concept enables asymmetric synthetic chemistry to set foot into the field of mobile supramolecular chemistry linked non-covalent system, beyond the conventional fixed covalent system. Furthermore, considering the high interest [40–42] in applications of supramolecules in areas as diverse as molecular machine [43, 44], material science [45–47], medicinal chemistry [48–50], and synthetic chemistry [51–55], I expect that attractive chiral architectures of mechanically chiral supramolecules will bring a novel chemical space into these fields, as well as in the field of organic synthesis.

References

1. Eder U, Sauer G, Wiechert R (1971) New type of asymmetric cyclization to optically active steroid CD partial structures. Angew Chem Int Edn 10:496–497
2. Hajos ZG, Parrish DR (1974) Asymmetric synthesis of bicyclic intermediates of natural product chemistry. J Org Chem 39:1615–1621
3. List B, Lerner RA, Barbas CF III (2000) Proline-catalyzed direct asymmetric aldol reactions. J Am Chem Soc 122:2395–2396

4. Notz W, List B (2000) Catalytic asymmetric synthesis of *anti*-1,2-diols. J Am Chem Soc 122:7386–7387
5. List B, Pojarliev P, Castello C (2001) Proline-catalyzed asymmetric aldolreactions between ketones and α-unsubstituted aldehydes. Org Lett 3:573–575
6. Jarvo ER, Miller SJ (2002) Amino acids and peptides as asymmetric organocatalysts. Tetrahedron 58:2481–2495
7. Forni A, Moretti I, Prosyanik AV, Torr G (1981) Optically active trifluoromethylcarbinols as chiral solvating agents for asymmetric transformations at a ring -nitrogen atom. J Chem Soc Chem Commun, 588–590
8. Christie GH, Kenner J (1922) The molecular configurations of polynuclear aromatic compounds. Part I. The resolution of γ-6:6'-Dinitro- and 4:6:4':6'-tetranitro-diphenic acids into optically active components. J Chem Soc Trans 121:614–620
9. Cope AC, Ganellin CR, Jhonson HW, Van Auken TV, Winkler HJS (1963) Molecular asymmetry of olefins. I. Resolution of *trans*-Cycloctene. J Am Chem Soc 85:3276–3279
10. Cope AC, Pawson BA (1965) Molecular asymmetry of olefins. IV. Kinetics of racemization of (+ or –)-*trans*-Cyclooctene. J Am Chem Soc 87:3649–3651
11. Newman MS, Lutz WB, Lednicer D (1955) A new reagent for resolution by complex formation; the resolution of Phenanthro-[3,4-*c*]Phenanethrene. J Am Chem Soc 77:3420–3421
12. Newman MS, Lednicer D (1956) The synthesis and resolution of hexahelicene. J Am Chem Soc 78:4765–4770
13. Gottarelli G, Hansen HJ, Spada GP, Weber RH (1987) 50. A liquid-crystal study of heptalene. Helvetica Chim Acta 70:430–435
14. Finocchiaro P, Gust D, Mislow K (1974) Structure and dynamic stereochemistry of trimesitylmethane. I. Synthesis and nuclear magnetic resonance studies. J Am Chem Soc 96:2165–2167
15. Andose JD, Mislow K (1974) Structure and dynamic stereochemistry of trimesitylmethane. 1. Empirical force field calculations. J Am Chem Soc 96:2168–2176
16. Finocchiaro P, Gust D, Mislow K (1947) Stereochemistry of Dimesityl-9-anthrylmethane and Bis(2,6-xylyl)-1-(2-methylnaphthyl)methane. Evidence for the two-ring flip mechanism in triarylmethanes. J Am Chem Soc 96:2176–2182
17. Hayes KS, Nagumo M, Blount JF, Mislow K (1980) Structure, optical resolution, and conformational stability of perchlorotriphenylamine. J Am Chem Soc 102:2773–2776
18. Thilgen C, Herrmann A, Diederieh F (1997) 16. Configurational description of chiral fullerenes and fullerene derivatives with a chiral functionalization pattern. Helvetica Chim Acta 80:183–199
19. Shirakawa S, Liu S, Kaneko S (2016) Organocatalyzed asymmetric synthesis of axially, planar, and helical chiral compounds. Chem Asian J 11:330–341
20. Cheng D-J, Yan L, Tian S-K, Wu M-Y, Wang L-X, Fan Z-L, Zheng S-C, Liu X-Y, Tan B (2014) Highly enantioselective kinetic resolution of axially chiral BINAM derivatives catalyzed by a brønsted acid. Angew Chem-Int Edn 53:3684–3687
21. Miyaji R, Asano K, Matsubara S (2015) Bifunctional organocatalysts for the enantioselective synthesis of axially chiral isoquinoline *N*-oxides. J Am Chem Soc 137:6766–6769
22. Link A, Sparr C (2014) Organocatalytic atroposelective aldol condensation: synthesis of axially chiral biaryls by arene formation. Angew Chem-Int Edn 53:5458–5461
23. Di Iorio N, Righi P, Mazzanti A, Mancinelli M, Ciogli A, Bencivenni G (2014) Remote control of axial chirality: aminocatalytic desymmetrization of *N*-arylmaleimides via vinylogous Michael addition. J Am Chem Soc 136:10250–10253
24. Kötzner L, Webber MJ, Martínez A, De Fusco C, List B (2014) Asymmetric catalysis on the nanoscale: the organocatalytic approach to helicenes. Angew Chem-Int Edn 53:5202–5205
25. Hawkins JM, Meyer A (1993) Optically active carbon: kinetic resolution of C_{76} by asymmetric osmylation. Science 260:1918–1920
26. Shoji Y, Tashiro K, Aida T (2010) One-pot enantioselective extraction of chiral fullerene C_{76} using a cyclic host carrying an asymmetrically distorted, highly π-basic porphyrin module. J Am Chem Soc 132:5928–5929

27. Ikeda A, Ishikawa M, Kikuchi J, Nobusawa K (2013) Enantioselective recognition of [60]Fullerene-Bisadducts by Cyclodextrin: kinetic control of a mechanochemical high-speed vibration milling process. Chem Lett 42:1137–1139
28. Pérez EM, Martín N (2012) Chiral recognition of carbon nanoforms. Org Biomol Chem 10:3577–3583
29. Bruns CJ, Stoddart JF (2016) The nature of the mechanical bond: from molecules to machines. Wiley
30. Frisch HL, Wasserman E (1961) Chemical topology. J Am Chem Soc 83:3789–3795
31. Liang C, Mislow K (1995) Topological chirality and achirality of links. J Math Chem 18:1–24
32. Wolf C (2008) Dynamic stereochemistry of chiral compounds: principles and applications. RSC (Royal Society of Chemistry) Publishing, Cambridge
33. Makita Y, Kihara N, Nakakoji N, Takata T, Inagaki S, Yamamoto C, Okamoto Y (2007) Catalytic asymmetric synthesis and optical resolution of planar chiral rotaxane. Chem Lett 36:162–163
34. Kawabata T, Nagato M, Takasu K, Fuji K (1997) Nonenzymatic kinetic resolution of racemic alcohols through an "Induced Fit" process. J Am Chem Soc 119:3169–3170
35. Schedel H, Kan K, Ueda Y, Mishiro K, Yoshida K, Furuta T, Kawabata T (2012) Asymmetric desymmetrization of meso-diols by C_2-symmetric chiral 4-pyrrolidinopyridines. Beilstein J Org Chem 8:1778–1787
36. Yoshida K, Furuta T, Kawabata T (2011) Organocatalytic chemoselective monoacylation of 1, n-linear diols. Angew Chem-Int Edn 50:4888–4892
37. Yoshida K, Shigeta T, Furuta T, Kawabata T (2012) Catalyst-controlled reversal of chemoselectivity in acylation of 2-aminopentane-1,5-diol derivatives. Chem Commun 48:6981–6983
38. Kawabata T, Muramatsu W, Nishio T, Shibata T, Schedel H (2007) A catalytic one-step process for the chemo- and regioselective acylation of monosaccharides. J Am Chem Soc 129:12890–12895
39. Yoshida K, Mishiro K, Ueda Y, Shigeta T, Furuta T, Kawabata T (2012) Non-enzymatic geometry-selective acylation of Tri- and tetrasubstituted d, dt-alkenediols. Adv Synth Catal 354:3291–3298
40. Erbas-Cakmak S, Leigh DA, McTernan CT, Nussbaumer AL (2015) Artificial molecular machines. Chem Rev 115:10081–10206
41. Evans NH (2018) Chiral catenanes and rotaxanes: fundamentals and emerging applications. Chem Eur J 24:3101–3112
42. Nakazono K, Takata T (2020) Mechanical chirality of rotaxanes: synthesis and function. Symmetry 12:144
43. Kay ER, Leigh DA, Zerbetto F (2007) Synthetic molecular motors and mechanical machines. Angew Chem Int Edn 46:72–191
44. Neal EA, Goldup SM (2014) Chemical consequences of mechanical bonding in catenanes and rotaxanes: isomerism, modification, catalysis and molecular machines for synthesis. Chem Commun 50:5128–5142
45. Ito K (2007) Novel cross-linking concept of polymer network: synthesis, structure, and properties of slide-ring gels with freely movable junctions. Polym J 39:489–499
46. Okumura Y, Ito K (2001) The Polyrotaxane gel: a topological gel by figure-of-eight cross-links. Adv Mater 13:485–487
47. David DHG, Casares R, Cuerva RJM, Campana AG, Blanco V (2019) A [2]Rotaxane-based circularly polarized luminescence switch. J Am Chem Soc 141:18064–18074
48. Hara K, Beppu T, Kimura M, Fujita Y, Takata T, Nishio K, Ono N (2013) Influence of novel supramolecular substance, [2]Rotaxane, on the caspase signaling pathway in Melanoma and colon cancer cells in vitro. J Pharm Sci 122:153–157
49. Fujita Y, Kimura M, Sato H, Takata T, Ono N, Nishio K (2016) Characterization of the cytotoxic activity of [2]Rotaxane (TROA0001), a novel supramolecular compound, in cancer cells. Arch Pharmacal Res 39:825–832
50. Barat R, Legigan T, Tranoy-Opalinski I, Renoux B, Péraudeau E, Clarhaut J, Poinot P, Fernandes AE, Aucagne V, Leigh DA, Papot S (2015) A mechanically interlocked molecular system programed for the delivery of an anticancer drug. Chem Sci 6:2608–2613

51. Jamieson EMG, Modicom F, Goldup SM (2018) Chirality in rotaxanes and catenanes. Chem Soc Rev 47:5266–5311
52. Leigh DA, Marcos V, Wilson MR (2014) Rotaxane catalysts. ACS Catal 4:4490–4497
53. Heard AW, Goldup SM (2020) Synthesis of a mechanically planar chiral rotaxane ligand for enantioselective catalysis. Chem 6:994–1006
54. Martinez-Cuezva A, Saura-Sanmartin A, Alajarin M, Berna J (2020) Mechanically interlocked catalysts for asymmetric synthesis. ACS Catal 10:7719–7733
55. Pairault N, Niemeyer J (2018) Chiral mechanically interlocked molecules—applications of rotaxanes, catenates and molecular knots in stereoselective chemosensing and catalysis. Synlett 29:689–698

Chapter 2
Mechanistic Study of Organocatalytic Chemoselective Monoacylation of 1,5-Pentanediol

Abstract Fine molecular transformation is one of the fundamental challenges in organic synthesis. Kawabata and co-workers have developed various acylative molecular transformations of polyol compounds by using C_2-symmetric chiral 4-pyrrolidinopyridine (PPY) catalysts bearing substrate-recognition sites consisting of amino acid side chains (Yoshida et al. in Angew Chem-Int Edn 50:4888–4892, 2011, [1]; Kawabata et al. in J Am Chem Soc 129:12890–12895, 2007, [2]; Yoshida et al. in Adv Synth Catal 354:3291–3298, 2012, [3]). Among them, I focused on the organocatalytic chemoselective monoacylation of 1,5-penetanediol (**2**) (Yoshida et al. in Angew Chem-Int Edn 50:4888–4892, 2011, [1]) for elucidation of the molecular recognition process promoted by the catalyst (Fig. 2.1a) (Kawabata et al. in J Am Chem Soc 119:3169–3170, 1997, [4]). To elucidate the origin of the high selectivity for monoacylation, I investigated the monoacylation of 1,5-pentanediol (**2**) in the presence of C_2-symmetric catalysts **1a–c**. The amide carbonyl group of the catalysts was suggested to play a main role for the selective monoacylation of **2**. The indolyl NH groups in the side chains of the catalysts seem to also contribute to increasing the chemoselectivity and reactivity. The effect of the indolyl NH groups were found to be significant only when the catalyst has C_2-symmetric structure by comparison with the experimental results employing the corresponding C_1-symmetric catalysts. Although 1,5-pentanediol (**2**) has only hydroxy groups as recognition sites, the multiple hydrogen bonding network between **2** and the catalyst was found to be essential for fine molecular recognition (Fig. 2.1b). Theoretical study was also well consistent with the experimental results. C_2-symmetric chiral PPY catalyst **1a** flexibly changes its molecular recognition mode depending on the substrate structures, thereby promoting chemoselective monoacylation of **2**.

Keywords Acylative molecular transformations · Mechanistic study · PPY catalysts · Chemoselective monoacylation

A. Imayoshi, *Discrimination of Mobile Supramolecular Chirality*, Springer Theses,
https://doi.org/10.1007/978-981-16-7431-0_2

2.1 Backgrounds

2.1.1 Backgrounds of Acylative Molecular Transformations

In our laboratory (Kawabata group), various molecular transformations concerning site-selective acylation of polyol compounds have been developed by using PPY catalysts bearing substrate-recognition sites consisting of amino acid side chains [1–3] (Figs. 2.2 and 2.3).

For example, in the presence of chiral PPY catalyst **1a** consisting of L-tryptophan side chains, the chemoselective monoacylation of 1,5-pentanediol (**2**) proceeded to give the monoacylate **3** and the diacylate **4** in 92% and 3% yield, respectively (**3**/**4** = 31) [1]. Catalyst **1a** also promoted substrate-selective monoacylation of 1,*n*-linear diols. 1,5-Pentanediol (**2**) was preferentially acylated 5.2 times faster than a one-carbon-longer diol (1,6-hexnaediol), in the competitive acylation between them (Fig. 2.2a). The observed phenomena obviously indicate that the molecular recognition process between catalyst **1a** and the substrate diols would be involved. Catalyst **1a** recognize the overall molecular structures of substrates to give selectively acylated products.

The chemo- and regioselective acylation of glycopyranoses have been developed with chiral PPY catalyst **1a** (Fig. 2.2b) [2]. Acylation of the secondary hydroxy group at C(4) of octyl β-D-glucopyranoside proceeded with >99% selectivity in the presence of a primary hydroxy group and two other secondary hydroxy groups. The site-selective acylation took place preferentially even in the presence of a primary alcohol (2-phenyletanol), which was supposed to be achieved by the precise molecular recognition promoted by catalyst **1a**.

Chiral PPY catalyst **1c,** bearing L-2-naphthylalanine side chains, has also been known to achieve various selective molecular transformations. The geometry-selective acylation of NsHN-substituted α,α′-alkenediols (Ns = 2-NO_2–$C_6H_4SO_2$–) have been achieved in >99% *E*-selectivity (Fig. 2.3a) [3].

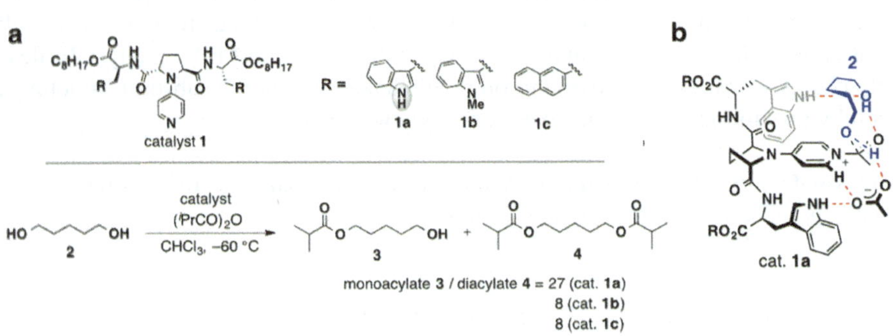

Fig. 2.1 Mechanistic study for organocatalytic chemoselective monoacylation of 1,5-pentanediol (**2**). **a** Chemoselective monoacylation of **2** catalyzed by **1a–1c**, **b** transition state of monoacylation of **2** catalyzed by **1a**

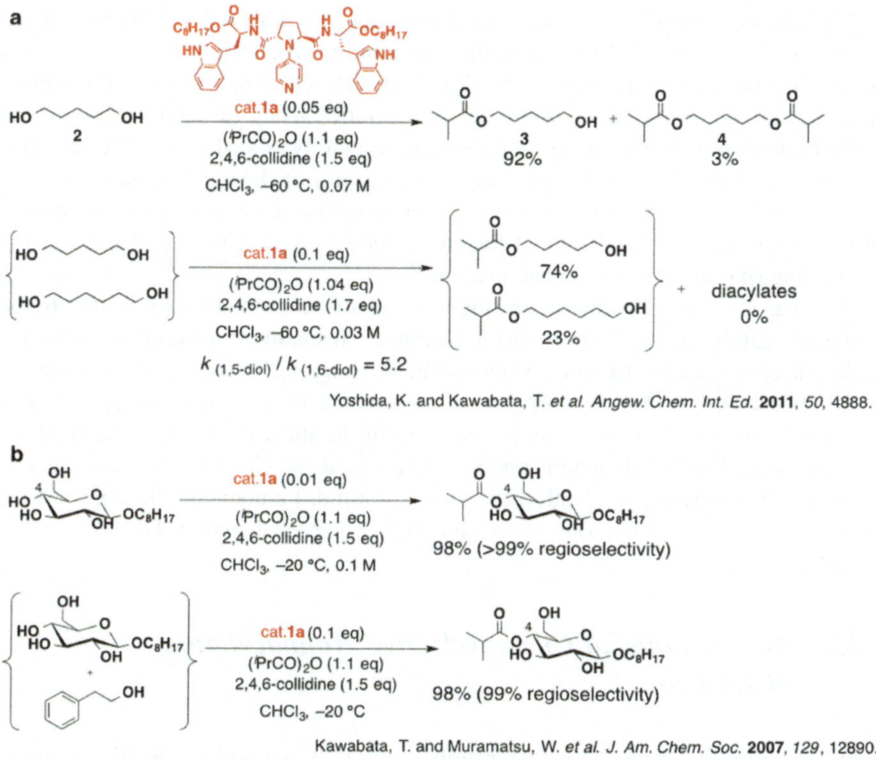

Fig. 2.2 Molecular transformations by chiral PPY catalyst **1a**. **a** Chemoselective acylation of linear diols and competitive acylation between diols having different chain length, **b** chemo- and regioselective acylation of glycopyranose and competitive acylation between glycopyranose and primary alcohol

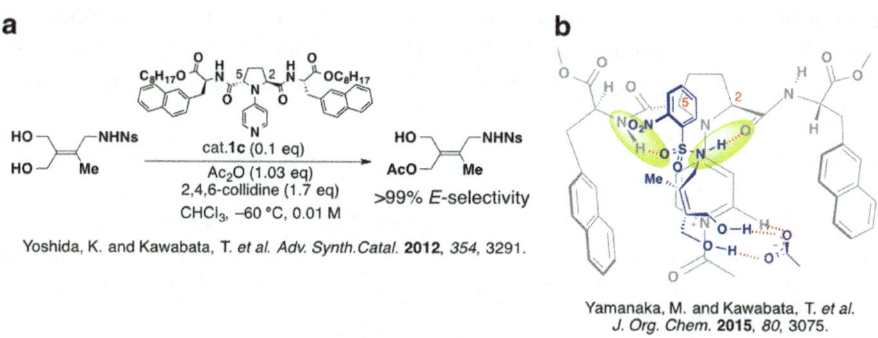

Fig. 2.3 Molecular transformations by chiral PPY catalyst **1c**. **a** Geometry-selective acylation of NsHN-substituted α,α'-alkenediols, **b** the transition state model

Mechanistic insights into the selective acylation catalyzed by **1c** have been proposed (Fig. 2.3b) [5]. Multiple hydrogen-bonding interactions between catalyst **1c** and the substrates play important roles in the molecular recognition. The fundamental hydrogen bond between the amide carbonyl group of catalyst **1c** and the acidic hydrogen of NHNs group in the substrate mainly contributes to the selective acylation, and simultaneously the other amide group of the catalyst acts cooperatively to further stabilize transition state of the selective acylation. Namely, catalyst **1c** possesses the dual functions of the two amide groups at C(2) and C(5) of the pyrrolidine ring in a C_2-symmetric fashion.

Whereas intriguing results suggesting mechanistic aspects of the molecular transformation catalyzed by **1c** have been obtained, mechanistic insights of selective acylation catalyzed by **1a** remain unexplained. Since the selective monoacylation of 1,5-diol **2** (Fig. 2.2a) [1] seems to be one of the most simple system because 1,5-diol **2** has only hydroxy groups as recognition sites, its mechanistic analysis is expected to lead fundamental understanding of these chemo- and site-selective molecular transformations. With these views in mind, I elucidated the origin of the high chemoselectivity on monoacylation of 1,5-diol **2** catalyzed by **1a**.

2.1.2 *Backgrounds of Chemoselective Monoacylation of 1,n-Diol*

Selective monoacylation of 1,*n*-linear diols seems to be a simple molecular transformation, however, it has been one of the difficult subject in organic synthesis because overacylation affording the diacylate is usually unavoidable, especially in the acylation of long-chain linear diols. For example, acylation of 1,5-pentanediol (**2**) with typical acylation catalyst, 4-dimethylaminopyridine (DMAP), generated significant amount of diacylate **4** as well as monoacylate **3** (**3**/**4** = 45%/26%), even under the carefully controlled conditions at −60 °C (Fig. 2.4, entry 1) [1]. The relative rate ($k_{1\ (\mathrm{acylation\ of\ 2})}/k_{2\ (\mathrm{acylation\ of\ 3})}$) for acylation between 1,5-diol **2** and monoacylate **3**

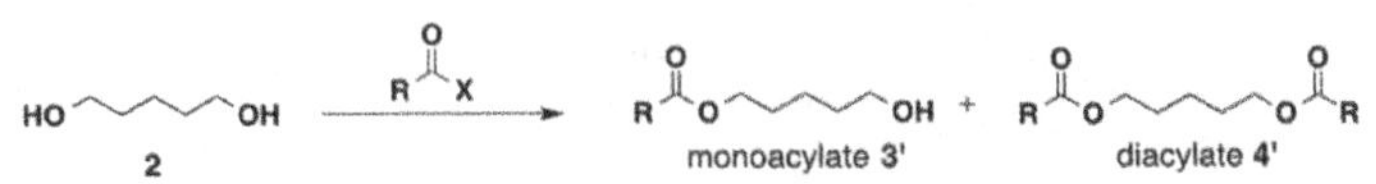

entry	conditions	R	monoacylate 3' (%)	diacylate 4' (%)	mono-/diacylate
1*	DMAP (0.05 eq), (^{i}PrCO)$_2$O, CHCl$_3$, −60 °C	^{i}Pr	45	26	1.7
2‡	DCC, ArSO$_2$CH$_2$CO$_2$H, THF	ArSO$_2$CH$_2$	70	20	3.5
3#	Novozyme 871, AcOCH=CH$_2$, toluene	Me	81	9	9.0
4*	**1a** (0.05 eq), (^{i}PrCO)$_2$O (1.1 eq), CHCl$_3$, −60 °C	^{i}Pr	92	3	31

Ar = benzothiazole-2-yl

* Yoshida, K. and Kawabata, T. *et al. Angew. Chem. Int. Ed.* **2011**, *50*, 4888.
‡ P. R. Blakemore *et al. Synlett* **2010**, 374.
\# P. Clapes *et al. Tetrahedron Lett.* **2004**, *45*, 5031.

Fig. 2.4 Mono- versus diacylation of 1,5-pentanediol (**2**)

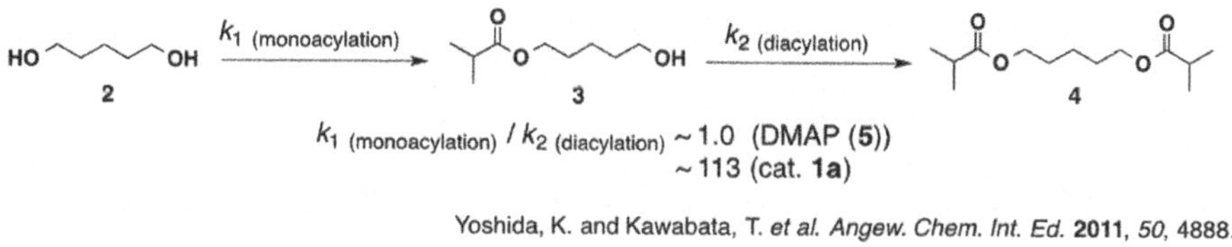

Fig. 2.5 Relative rate for acylation ($k_{1\,\text{(monoacylation)}}/k_{2\,\text{(diacylation)}}$) between diol **2** and monool **3**

with DMAP catalyst was experimentally estimated to be 1.0 (Fig. 2.5) [1]. This result indicates that the intrinsic reactivity of 1,5-diol **2** was assumed to be equal to that of monoacylate **3**.

To avoid overacylation [6], excess amounts of the diol substrates were often used [7–9]. For example, acylation of 3.0–6.0 equivalents of **2** with a carboxylic acid in the presence of *N*,*N′*-dicyclohexylcarbodiimide (DCC) gave the monoacylate and the diacylate in a 3.5: 1 ratio (Fig. 2.4, entry 2) [9]. Selective acylation of **2** has been reported by the use of an enzyme [10, 11], however, overacylation is again unavoidable (mono/diacylate = 9.0, entry 3) [11]. Under these backgrounds, in our laboratory, highly selective monoacylation of 1,*n*-linear diols have been achieved by using chiral PPY catalyst **1a** (entry 4) [1]. In the presence of catalyst **1a**, acylation of 1,5-diol **2** gave the monoacylate **3** in a highly chemoselective manner (mono/diacylate = 31).

For such highly selective monoacylation catalyzed by **1a**, the step for monoacylation must be much faster than the step for diacylation ($k_{1\,\text{(monoacylation)}} \gg k_{2\,\text{(diacylation)}}$). A high relative rate ($k_{1\,\text{(monoacylation)}}/k_{2\,\text{(diacylation)}}$) is supposed to be essential for selective monoacylation. In fact, the relative rate ($k_{1\,\text{(monoacylation)}}/k_{2\,\text{(diacylation)}}$) for acylation between 1,5-diol **2** and monoacylate **3** with catalyst **1a** was experimentally estimated to be 113 (Fig. 2.5) [1]. The strong preference for acylation of **2** over that of **3** in the presence of **1a** indicates the precise molecular recognition of the substrate structure by catalyst **1a** on the selective monoacylation.

The preferential acylation of 1,5-diol **2** over other linear diols with different chain length has been also achieved in the presence of catalyst **1a** [1]. The relative rates ($k_{\text{(acylation of 1,}n\text{-diol)}}/k_{\text{(acylation of 1,5-diol 2)}}$) of acylation between 1,*n*-diol ($n = 4$, 6 or 7) and 1,5-diol **2** are shown in Fig. 2.6. The acylation of 1,5-diol **2** proceeded 2.0, 5.6, and 6.6 times faster than that of 1,4-butanediol, 1,6-hexanediol, and 1,7-heptanediol, respectively (Fig. 2.6a). On the other hand, negligible substrate-selectivity was observed in DMAP-catalyzed acylation of these diols (Fig. 2.6b). Thus, catalyst **1a** was found to promote substrate-selective monoacylation of 1,*n*-linear diols by discriminating the just one-carbon difference in their chain length.

The hypothetical transition state model for monoacylation of **2** catalyzed by **1a** (Fig. 2.7) has been proposed on the basis of some experimental results (Fig. 2.8) [1]. Hydrogen-bonding between catalyst **1a** and the primary hydroxy group of 1,5-diol **2** is assumed to be essential for selective monoacylation from the experimentally observed solvent and temperature effects (Fig. 2.8a, b). Two amide carbonyl groups

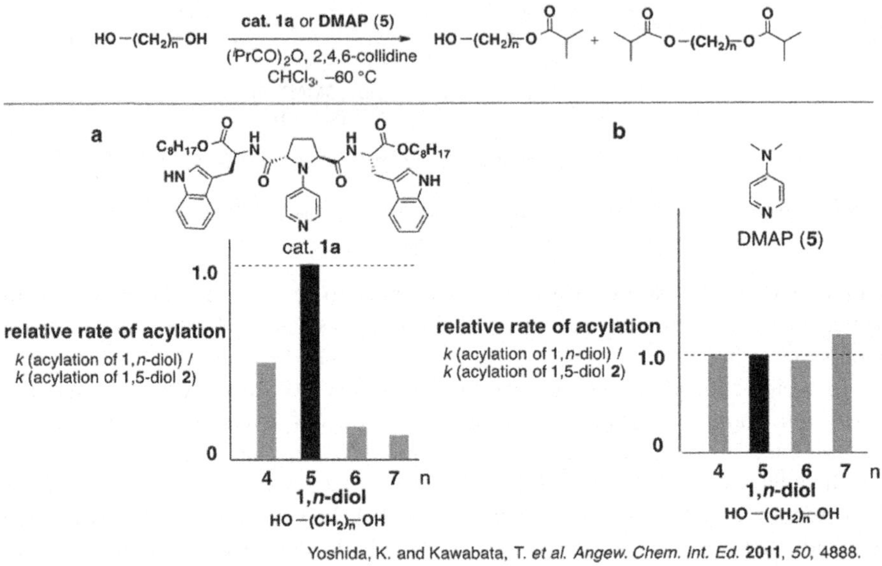

Fig. 2.6 Relative rate of acylation ($k_{\text{(acylation of 1,}n\text{-diol)}}/k_{\text{(acylation of 1,5-diol 2)}}$). **a** Catalyzed by **1a**, **b** catalyzed by DMAP (**5**)

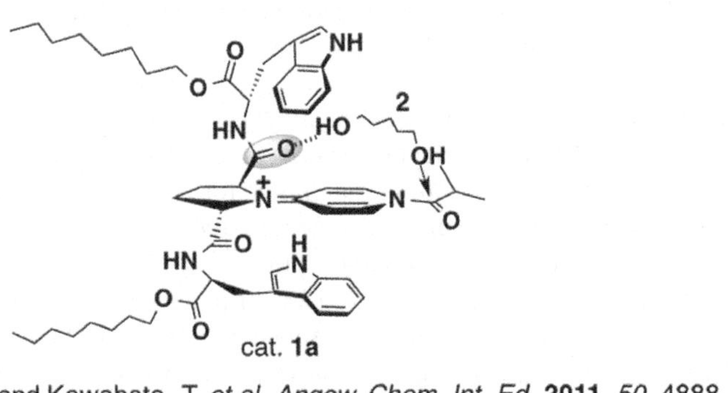

Fig. 2.7 Previously proposed hypothetical transition state model without theoretical base for preferential monoacylation of 1,5-diol **2** catalyzed by **1a**

seem to be important for the chemoselective acylation because selective monoacylation took place throughout the reactions with catalysts **1a**, **1d** and **1e**, each possessing two amide carbonyl groups as common functionalities, regardless of the steric bulkiness of the side chains (Fig. 2.8c). The model shown in Fig. 2.7 may account for the preferential acylation of 1,5-diol **2** over the corresponding monoacylate **3** and substrate-specific acylation of 1,5-diol **2**. However, the actual mechanism seems to be not so simple. Since structures of catalyst side chains affect the degree of

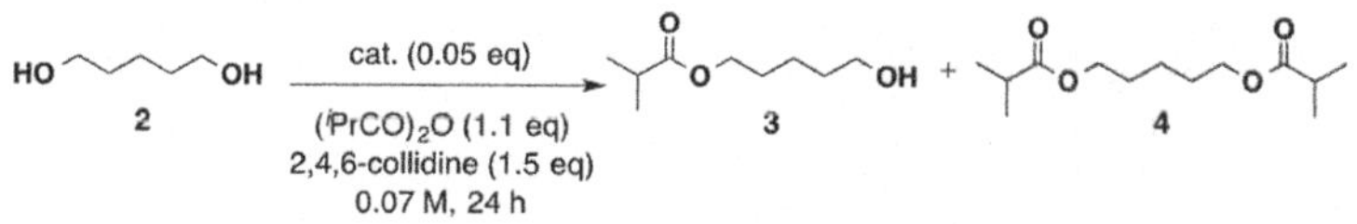

a

temp. (°C)	3/4
−60	27
−20	9
20	6

b

solvent	3/4
$CHCl_3$	>71
toluene	8
DMF	2

c

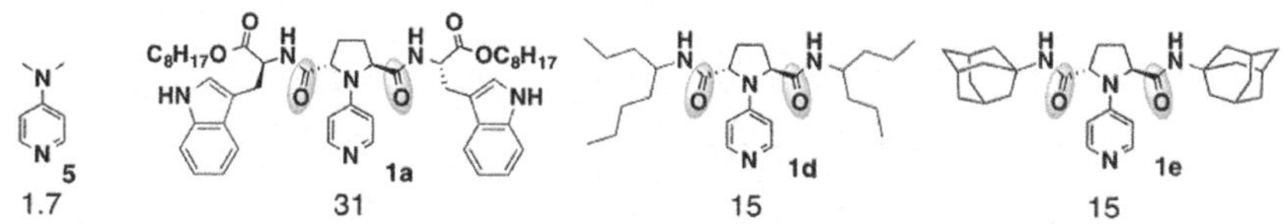

3/4	1.7	31	15	15

Yoshida, K. and Kawabata, T. *et al.* *Angew. Chem. Int. Ed.* **2011**, *50*, 4888.

Fig. 2.8 Various effects on selective monoacylation of 1,5-diol **2**. **a** Effects of temperature (Run with catalyst **1a** in $CHCl_3$), **b** effects of solvents (Run with catalyst **1a** at −60 °C), **c** effects of steric bulk of catalyst side chains (Run in $CHCl_3$ at −60 °C)

selectivity (Fig. 2.8c), the catalyst side chains seem to play important roles for the chemoselectivity. These backgrounds prompted me to further elucidate the mechanistic aspects of the organocatalytic chemoselective acylation based on experimental and computational studies.

2.2 Experimental Studies for Elucidation of Mechanistic Aspects of Chemoselective Monoacylation of 1,5-Pentanediol

2.2.1 Effects of Side Chain Structures of Catalysts

I first investigated the effects of the side chain structures of catalysts on chemoselective monoacylation of **2** (Fig. 2.9). I performed the acylation of 1,5-diol **2** with 1.03 equivalents of isobutyric anhydride in the presence of C_2-symmetric chiral PPY catalysts **1a–c**, each having two identical side chains consisting of L-tryptophan, 1-mehyl-L-tryptophan and L-2-naphthylalanine moiety, respectively. Among these catalysts, the distinct high selectivity for monoacylation was observed with catalyst **1a** (**3/4** = 27, entry 1). With catalyst **1b** or **1c**, monoacylate was still predominant, but with decreased chemoselectivity (**3/4** = 8) and reactivity (57–63% conversion, entries 2, 3), when compared with the acylation promoted by catalyst **1a** (**3/4** = 27,

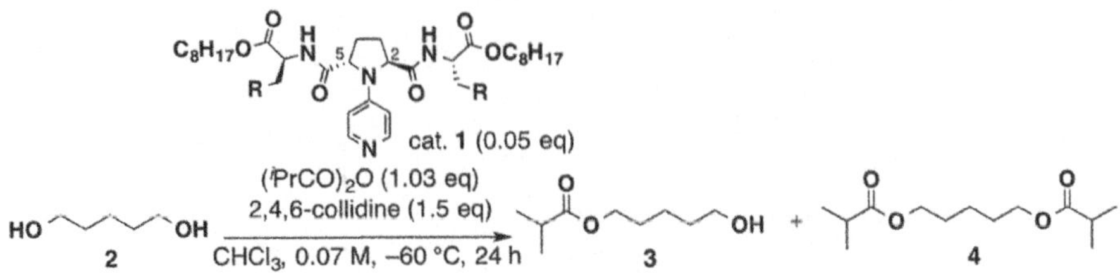

entry	R		3 : 4 (%) (%)	3/4	Recovery of **2** (%)
1	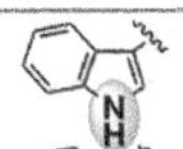	cat. **1a**	80 : 3	27	16
2	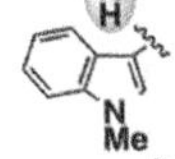	cat. **1b**	56 : 7	8	37
3	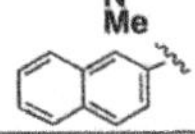	cat. **1c**	51 : 6	8	43

Fig. 2.9 Effects of side chains of catalysts **1** on chemoselective monoacylation of 1,5-diol **2**

84% conversion, entry 1). These results indicate the critical importance of the indolyl NH groups in catalyst **1a** for the chemoselective acylation.

2.2.2 *Effects of* C_2*-Symmetric Structures in Catalysts*

The effects of the C_2-symmetric structure of catalysts were next examined by comparing with the reaction using the corresponding C_1-symmetric catalysts **6a–c**. Almost no differences in chemoselectivity and reactivity were found among reactions catalyzed by C_1-symmetric **6a–c** (**3**/**4** = 10–13, 32% conversion, Fig. 2.10), whether the indolyl NH group was involved in catalyst structures **6a–c** or not. It is worthy to note that the indolyl NH group is responsible for further increasing the selectivity and reactivity in the monoacylation of 1,5-diol **2** *only when the PPY-catalyst has* C_2*-symmeric structure* (Figs. 2.9 vs. 2.10).

In the absence of indolyl NH group in the catalyst structure, C_1-symmetric catalysts **6b** and **6c** showed similar selectivity (**3**/**4** = 10–13) and diminished reactivity (32% conversion) compared to C_2-symmetric catalysts **1b** and **1c** (**3**/**4** = 8, 57–63% conversion). This result indicates the importance of the amide carbonyl groups for the chemoselectivity and reactivity, and the amide groups seem to contribute to acceleration of reactions. Since C_2-symmetric catalysts **1b** and **1c** have two amide groups, the reactivity was improved compared to the corresponding C_1-symmetric catalysts **6b** and **6c**, without affecting the chemoselectivity.

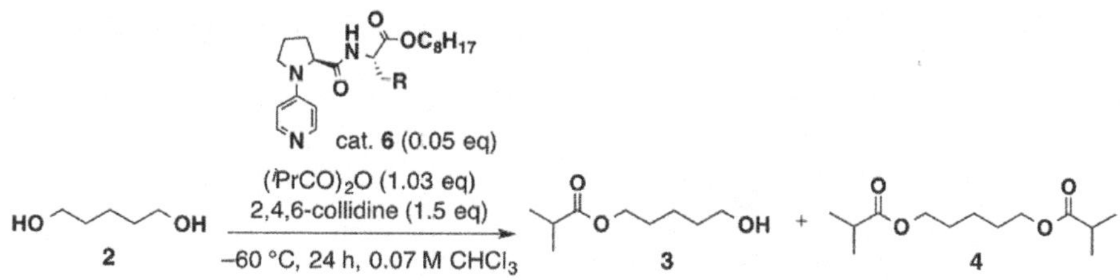

entry	R		3 : 4 (%) (%)	3/4	Recovery of 2 (%)
1	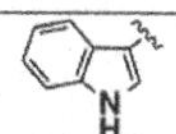	cat. **6a**	31 : 3	12	68
2	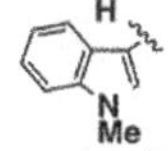	cat. **6b**	29 : 3	10	68
3	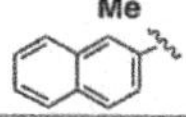	cat. **6c**	30 : 2	13	68

Fig. 2.10 Selective monoacylation of 1,5-diol **2** with C_1-symmetric catalysts **6**

2.3 Experimental Studies for Elucidation of the Mechanistic Aspects in Substrate-Selective Acylation of 1,5-Pentanediol

2.3.1 Effects of Catalyst Structures on Monoacylation of 1,2-Ethanediol

To elucidate the mechanistic aspects of substrate-selective acylation of 1,5-pentanediol (**2**), I next examined catalytic properties of **1a–c**, and **6a–c** in acylation of 1,2-ethanediol (**7**) (Fig. 2.11). C_2-symmetric PPY catalysts **1a–c** promoted chemoselective monoacylation of **7** in high selectivities (**8/9** = 23–35, Fig. 2.11a), whether or not the indolyl NH groups are involved in the catalyst structure. Indolyl NH groups of catalyst **1a** did not affect the chemoselectivity on monoacylation of 1,2-diol **7**, even though catalyst **1a** has C_2-symmetric structure. Thus, on acylation of 1,2-diol **7**, the indolyl NH groups have almost no impact on chemoselectivity. This result is a contrast to monoacylation of 1,5-diol **2**, where indolyl NH groups of C_2-symmetric catalyst **1a** played an important role for selective monoacylation (Fig. 2.9).

Chemoselective monoacylation of 1,2-diol **7** with C_1-symmetric catalysts **6a–c** proceeded in slightly less but similar selectivity and reactivity (**8/9** = 62%/5%–64%/4%, 64–68% conversion, Fig. 2.11b), when compared with the acylation promoted by C_2-symmetric catalyst structure (**8/9** = 69%/3%–70%/2%, 72–73% conversion, Fig. 2.11a). Thus, on acylation of 1,2-diol **7**, the amide carbonyl group at C(2) and/or C(5) of the pyrrolidine ring of the catalyst, as well as the indolyl NH group, hardly makes an impact on chemoselectivity and reactivity.

a

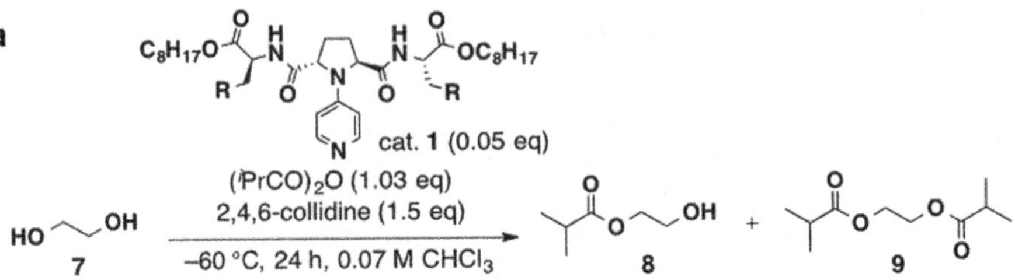

entry	R		8 : 9 (%) (%)	8/9	Recovery of 7 (%)
1	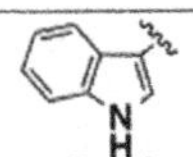	cat. **1a**	70 : 2	35	27
2	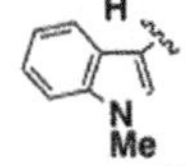	cat. **1b**	69 : 3	23	27
3	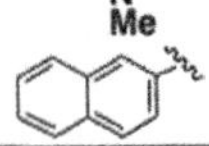	cat. **1c**	69 : 2	31	28

b

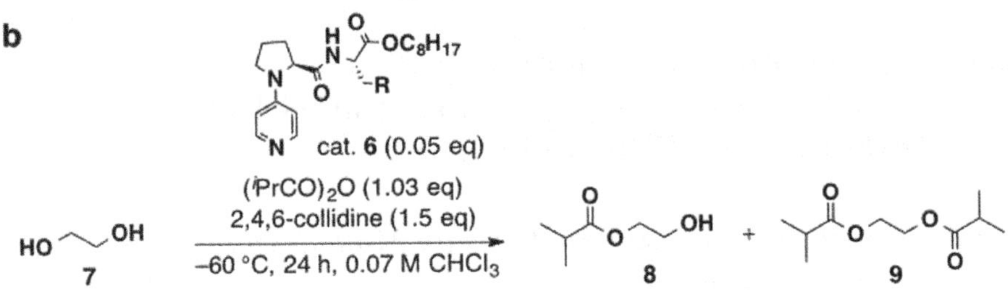

entry	R		8 : 9 (%) (%)	8/9	Recovery of 7 (%)
1	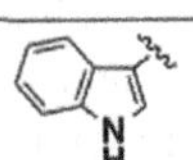	cat. **6a**	64 : 4	15	32
2	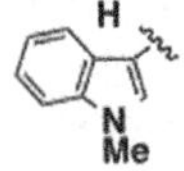	cat. **6b**	60 : 4	14	36
3	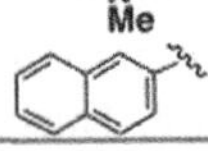	cat. **6c**	62 : 5	13	33

Fig. 2.11 Acylation of 1,2-ethanediol (**7**) by chiral PPY catalysts. **a** Catalyzed by C_2-symmetric **1a–1c**, **b** catalyzed by C_1-symmetric **6a–6c**

The acylation of 1,2-diol **7** showed relatively high chemoselectivity and reactivity throughout the reactions promoted by both C_1- and C_2-symmetric PPY-catalysts. This could be ascribed mainly to the less intrinsic reactivity of monool **8** than 1,2-diol **7**. Bulkiness of the acyl group of monoacylate **8** effectively decelerate the second acylation, especially in the cases of such short-chain diol, by using steric interaction to prevent the diacylation. 1,2-Diol **7** is also expected to have an intrinsically high reactivity for monoacylation because of its intramolecular hydrogen bonding.

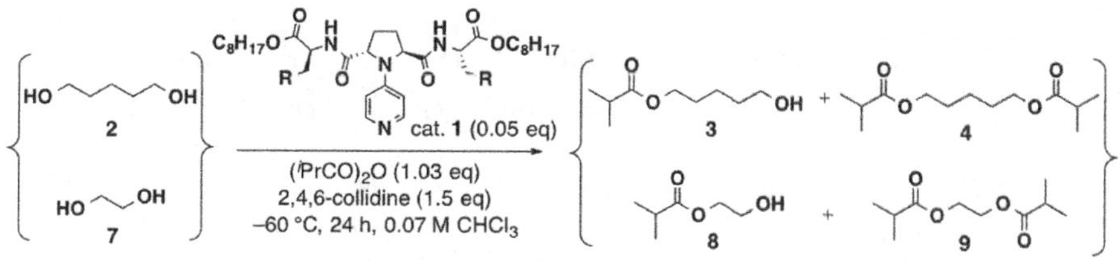

entry	R		3+4 : 8+9 (%) (%)	Chemoselectivity k (acylation of 1,5-diol **2**) / k (acylation of 1,2-diol **7**)
1	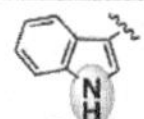	cat. **1a**	36 : 12	4.5
2	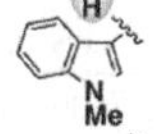	cat. **1b**	11 : 16	0.7
3	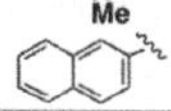	cat. **1c**	14 : 18	0.8

Fig. 2.12 Competitive acylation between 1,5-diol **2** and 1,2-diol **7** with catalysts **1a**–**1c**

2.3.2 *Effects of Catalyst Structures on Substrate-Selective Acylation of 1,5-Pentanediol*

To elucidate the mechanistic aspects of substrate-selective acylation of 1,5-pentanediol (**2**), relative rates of acylation (k (acylation of **2**)/k (acylation of **7**)) between 1,5-diol **2** and 1,2-diol **7** were determined by the competitive acylation between them in the presence of C_2-symmetric catalysts **1a**–**c** (Fig. 2.12). Catalyst **1a** with the indolyl NH groups promoted the acylation of 1,5-diol **2** 4.5 times faster than that of 1,2-diol **7** (entry 1). In contrast, in the presence of catalysts **1b** or **1c** without the indolyl NH group, 1,2-diol **7** was acylated relatively faster than 1,5-diol **2** (k (acylation of **2**)/k (acylation of **7** = 0.7–0.8, entries 2, 3). The marked difference in the catalytic properties between catalysts **1a**–**c** depending on the presence or the absence of the indolyl NH group indicates that two indolyl NH groups arranged in a C_2-symmetric fashion are essential for transition state stabilization, resulting in accelerative monoacylation of 1,5-diol **2**. The critical stabilization of transition state by indolyl NH groups enables 1,5-diol **2** to be preferentially acylated over the other linear diols.

2.4 Computational Study

Based on the experimental studies, theoretical study was performed by Professor Masahiro Yamanaka and Dr. Makoto Sato (Rikkyo University). DFT calculations (B3LYP/6-31G*, Gaussian 09 package) were carried out using C_1- and C_2-symmetric catalyst models (**cat6a** and **cat1a**, Fig. 2.13).

Transition state (TS) of the nucleophilic attack process in monoacylation of 1,5-diol **2** was investigated. The difference in relative Gibbs free energy between TSs of

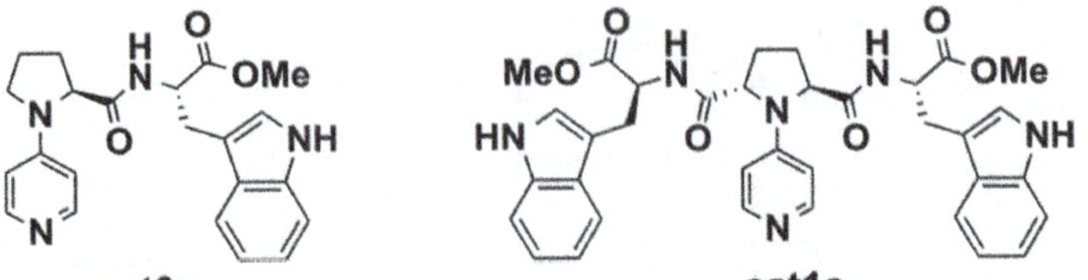

Fig. 2.13 Chemical models of catalysts **6a** and **1a**

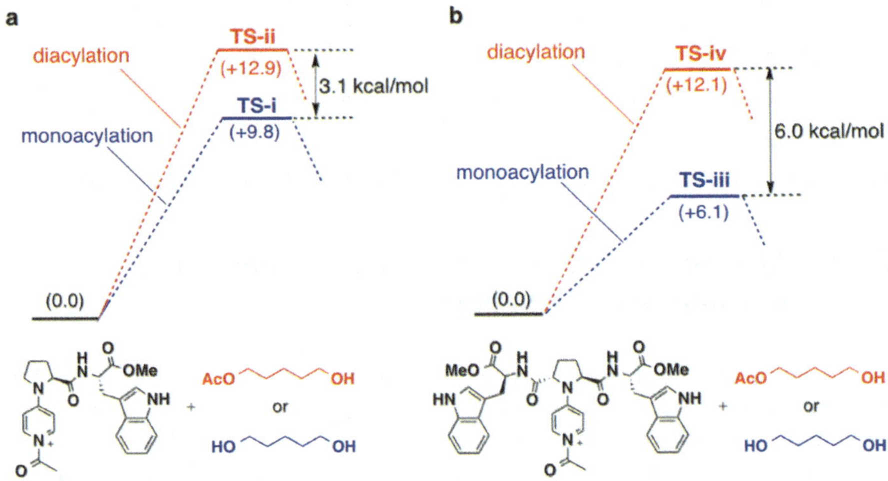

Fig. 2.14 Relative Gibbs free energy differences (kcal/mol) between transition states of monoacylation (blue) and diacylation (red). **a** Cat**6a** system, **b** Cat**1a** system (Calculated by Professor Masahiro Yamanaka and Dr. Makoto Sato (Rikkyo University) at B3LYP/6-31G* level)

monoacylation and diacylation was compared in each of the C_1-symmetric **cat6a** and C_2-symmetric **cat1a** systems (Fig. 2.14). Monoacylation is energetically favored than diacylation in both systems. Although the energy level of diacylation TSs (**TSii**, **TS-iv**) are almost the same in both systems, the significant stabilization of the monoacylation TS was observed in the **cat1a** system (**TS-iii**, Fig. 2.14b). These computational results are well consistent with the experimental results.

Figure 2.15 shows the most stable TSs in the C_1-symmetric **cat6a**-catalyzed monoacylation (**TS-i**, Fig. 2.15a) and diacylation (**TS-ii**, Fig. 2.15b). In both reactions, the proton abstraction from the nucleophilic hydroxy group by the acetate anion is observed [12, 13]. The acetate anion ($CO_{acetate}$) strongly coordinates with the indolyl NH group (NH_{indole}).

The major difference between **TS-i** and **TS-ii** is derived from the presence or absence of a hydrogen bond between the amide carbonyl group of **cat6a** (CO_{amide}) and the substrate. The energetically favored **TS-i** has the hydrogen bond between the amide carbonyl group of **cat6a** (CO_{amide}) and the remaining hydroxy group of

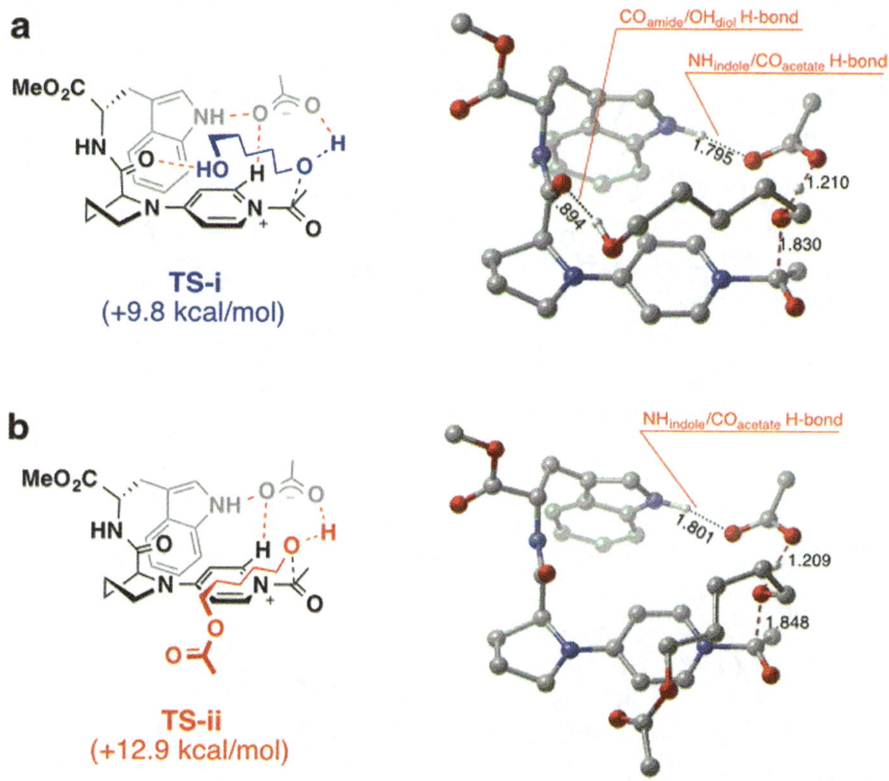

Fig. 2.15 Transition states in C_1-symmetric cat**6a** system. **a** TS of monoacylation (**TS-i**), **b** TS of diacylation (**TS-ii**) (Calculated by Professor Masahiro Yamanaka and Dr. Makoto Sato (Rikkyo University) at B3LYP/6-31G* level)

1,5-diol **2** (OH_{diol}). In contrast, **TS-ii** does not have such additional strong hydrogen bond. Thus, **TS-i** is located at the lower energy level compared to **TS-ii**. These computational results are well consistent with the experimental results shown in Fig. 2.10.

Then, TSs of monoacylation (**TS-iii**, Fig. 2.16a) and diacylation (**TS-iv**, Fig. 2.16b) in the **cat1a** system were explored. Both TSs have the strong $NH_{indole}/CO_{acetate}$ hydrogen bond in a similar manner to the **cat6a** system.

On the other hand, the second indolyl NH group (NH_{indole}) provides a significant impact on the stability of the monoacylation TS (**TS-iii**) (Fig. 2.16a). The remaining hydroxy group works as a general acid catalyst by coordinating with the *N*-acylpyridinium carbonyl oxygen (CO_{acyl}). The increased Lewis basicity of the oxygen atom of diol (OH_{diol}) forms the strong hydrogen bond with the second indolyl NH group (NH_{indole}/OH_{diol}). In this way, the most stable TS structure (**TS-iii**) is formed. **TS-v** (Fig. 2.17), having a similar coordination mode with **TS-i** (Fig. 2.15a), is no longer the most stable TS because the second indolyl group does not contribute to

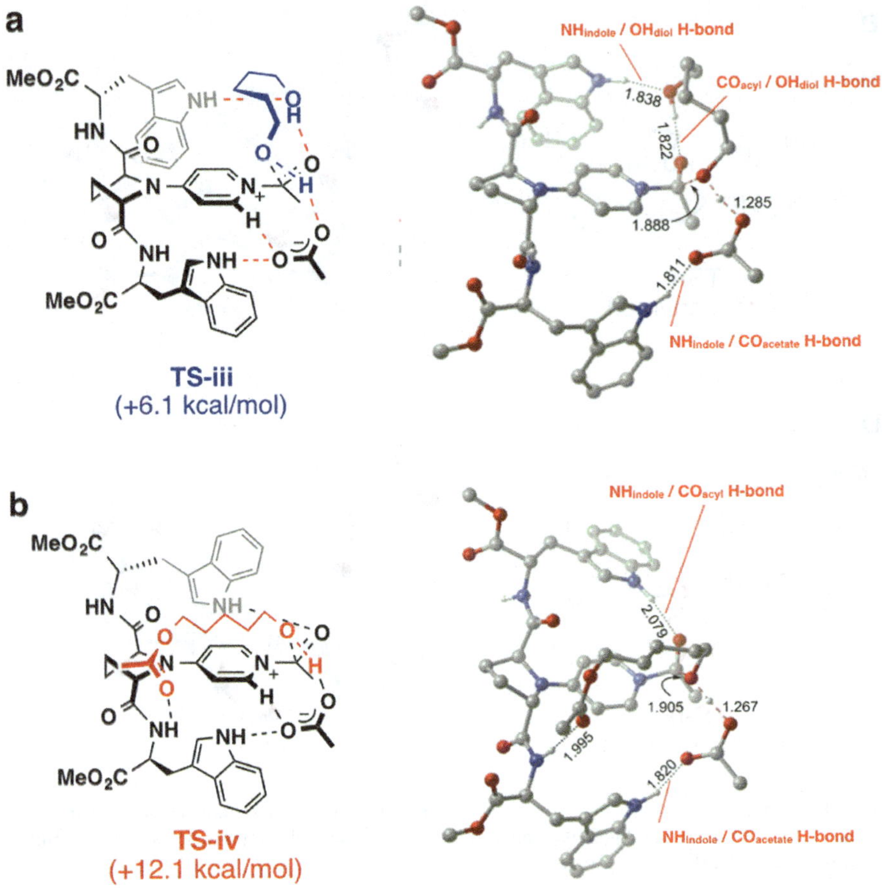

Fig. 2.16 Transition states in C_2-symmetric cat**1a** system. **a** TS of monoacylation (**TS-iii**), **b** TS of diacylation (**TS-iv**) (Calculated by Professor Masahiro Yamanaka and Dr. Makoto Sato (Rikkyo University) at B3LYP/6-31G* level)

the stabilization of **TS-v**. These results suggest that catalyst **1a** achieves the fine molecular recognition for 1,5-diol **2** by dramatically changing its coordination mode from catalyst **6a**.

In the diacylation TS of the **cat1a** system (**TS-iv**, Fig. 2.16b), the second NH_{indole} group directly coordinates with the *N*-acylpyridinium carbonyl oxygen (CO_{acyl}), forming the weak hydrogen bond. The second indolyl NH groups (NH_{indole}) rarely contribute to the hydrogen bonding network in the diacylation **TS-iv**.

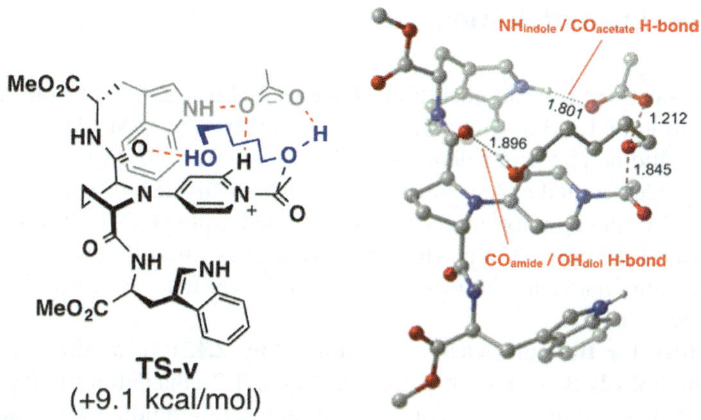

Fig. 2.17 Transition state in C_2-symmetric cat**1a** system without involving the second indolyl group (Calculated by Professor Masahiro Yamanaka and Dr. Makoto Sato (Rikkyo University) at B3LYP/6-31G* level)

2.5 Conclusion

Mechanistic aspects of chemoselective monoacylation of 1,5-pentanediol (**2**) has been experimentally and computationally investigated by employing various C_1- and C_2-symmetric PPY catalysts. Although 1,5-diol **2** has a simple structure possessing only hydroxy groups as recognition sites, the multiple hydrogen bonds between the C_2-symmetric catalyst **1a** and 1,5-diol **2** significantly stabilized the monoacylation transition state of 1,5-diol **2**. The indolyl NH groups arranged in a C_2-symmetric fashion were found to be critically important for both chemoselective monoacylation and the substrate-specific acylation of 1,5-diol **2**. C_2-symmetric catalyst **1a** performed the precise molecular recognition for 1,5-diol **2** through the quite different coordination mode from that by C_1-symmetric catalyst **6a**. Catalyst **1a** possesses the dual functions in the two indolyl NH groups of the L-tryptophan side chains.

Such recognition mode is quite different from that proposed for other molecular transformations as shown in Fig. 2.3. Thus, these chiral PPY catalysts seem to adapt themselves to various transformations *by flexibly changing their molecular recognition mode depending on the substrate structures*. These results prompted me to apply such molecular recognition modes by chiral PPY catalysts to more complex and flexible system. The topic for unprecedented asymmetric molecular transformations is discussed in Chap. 3.

2.6 Experimental Section

General procedure for the acylation of 1,*n*-diol (Figs. 2.9, 2.10 and 2.11) To a stirred solution of 1,*n*-diol (0.2 mmol, 1.0 equivalents), catalyst (0.05 equivalents) and 2,4,6-collidine (1.5 equivalents) in $CHCl_3$ (2.9 mL, 0.07 M for 1,*n*-diol) at −60 °C was added cooled isobutyric anhydride (1.03 equivalents). After stirring at −60 °C for 24 h, the reaction mixture was quenched with MeOH (10 mL), and the solution was stirred for 10 min. Then the solvent was carefully evaporated. Yields of monoacylate, diacylate and recovered 1,*n*-diol were determined by ^{1}H NMR in $CDCl_3$ or benzene-d_6.

Procedure for the competitive acylation (Fig. 2.12) To a stirred solution of 1,5-pentanediol (**2**) (0.16 mmol, 0.5 equivalents), 1,2-ethanediol (**7**) (0.16 mmol, 0.5 equivalents), catalyst (0.05 equivalents) and 2,4,6-collidine (1.5 equivalents) in $CHCl_3$ (4.6 mL, 0.07 M for the total amount of diols) at −60 °C was added cooled isobutyric anhydride (1.03 equivalents). After stirring at −60 °C for 24 h, the reaction mixture was quenched with MeOH (16 mL), and the solution was stirred for 10 min. Then the solvent was carefully evaporated. Yields of monoacylates, diacylates and recovered diols were determined by NMR in $CDCl_3$ and benzene-d_6. Chemoselectivity of the acylation of 1,5-pentanediol (**2**) versus 1,2-ethanediol (**7**) ($k_{\text{(acylation of 2)}}/k_{\text{(acylation of 7)}}$) were determined according to the following equation: $k_{\text{(acylation of 2)}}/k_{\text{(acylation of 7)}} = \ln[(1\text{–conversion})(1-|(\mathbf{8}+\mathbf{9})-(\mathbf{3}+\mathbf{4})|/\{(\mathbf{8}+\mathbf{9})+(\mathbf{3}+\mathbf{4})\})/\ln[(1\text{–conversion})(1+|(\mathbf{8}+\mathbf{9})-(\mathbf{3}+\mathbf{4})|/\{(\mathbf{8}+\mathbf{9})+(\mathbf{3}+\mathbf{4})\})]$. Conversion was calculated based on the total amount of the two diols.

References

1. Yoshida K, Furuta T, Kawabata T (2011) Organocatalytic chemoselective monoacylation of 1, n-linear diols. Angew Chem-Int Edn 50:4888–4892
2. Kawabata T, Muramatsu W, Nishio T, Shibata T, Schedel H (2007) A catalytic one-step process for the chemo- and regioselective acylation of monosaccharides. J Am Chem Soc 129:12890–12895
3. Yoshida K, Mishiro K, Ueda Y, Shigeta T, Furuta T, Kawabata T (2012) Non-enzymatic geometry-selective acylation of Tri- and tetrasubstituted α, α′-alkenediols. Adv Synth Catal 354:3291–3298
4. Kawabata T, Nagato M, Takasu K, Fuji K (1997) Nonenzymatic kinetic resolution of racemic alcohols through an "Induced fit" process. J Am Chem Soc 119:3169–3170
5. Yamanaka M, Yoshida U, Sato M, Shigeta T, Yoshida K, Furuta T, Kawabata T (2015) Origin of high *E*-selectivity in 4-pyrrolidinopyridine-catalyzed tetrasubstituted α, α′-alkenediols: a computational and experimental study. J Org Chem 80:3075–3082
6. Lewis CA, Sculimbrene BR, Xu Y, Miller SJ (2005) Desymmetrization of glycerol derivatives with peptide-based acylation catalysts. Org Lett 7:3021–3023
7. Takahashi S, Katagiri T, Uneyama K (2005) A binary hydrogen bonding motif based on homochiral recognition: crystal structures and hydrogen bonding networks of *meso*-(*R*,*S*)-bis(trifluorolactate)s. Chem Commun 3658–3660
8. Iwashita M, Makide K, Nonomura T, Misumi Y, Otani Y, Ishida M, Taguchi R, Tsujimoto M, Aoki J, Arai H, Ohwada T (2009) Synthesis and evaluation of lysophosphatidylserine analogues

as inducers of mast cell degranulation. Potent activities of lysophosphatidylthreonine and its 2-deoxy derivative. J Med Chem 52:5837–5863
9. Giesbrecht HE, Knight BJ, Tanguileg NR, Emerson CR, Blakemore PR (2010) Thieme chemistry journal awardees—where are they now? Stereoselective synthesis of Z-configured a, b-unsaturated macrocyclic lactones and diolides by intramolecular Julia-Kocienski Olefination. Synlett 3:374–378
10. Hajjar AB, Nicks PF, Knowles CJ (1990) Preparation of monomeric acrylic ester intermediates using lipase catalyzed transesterifications in organic solvents. Biotech Lett 12:825–830
11. Framis V, Camps F, Clapés P (2004) Lipase-catalysed selective monoacylation of 1, n-diols with vinyl acetate. Tetrahedron Lett 45:5031–5033
12. Xu S, Held I, Kempf B, Mayr H, Steglich W, Zipse H (2005) The DMAP-catalyzed acetylation of alcohols—a mechanistic study (DMAP=4-(Dimethylamino)pyridine). Chem Eur J 11:4751–4757
13. Larionov E, Zipse H (2011) Organocatalysis: acylation catalysts. Wiley Interdisc Rev Comput Mol Sci 1:601–619

Chapter 3
Discrimination of Mobile Supramolecular Chirality: Kinetic Resolution of Mechanically Planar Chiral Rotaxanes by Organocatalysis

Abstract Among the fields of asymmetric synthesis approaching the mature science, asymmetric discrimination and catalytic synthesis of chiral supramolecules still stand as unsolved problems. Supramolecules such as rotaxanes and catenanes are known to possess mechanical chirality when each of the axis and/or ring components has dissymmetry (Fig. 3.1) (Frisch and Wasserman in J Am Chem Soc 83:3789–3795, 1961, [1]; Schill in Catenanes, rotaxanes and knots, Academic Press, New York, 1971, [2]). The extreme difficulty in asymmetric synthesis of such supramolecules may be resulting from conformational diversity and movability of mechanically chiral supramolecules. I have achieved the first example of highly enantioselective synthesis of mechanically planar chiral rotaxanes by acylative kinetic resolution of the racemate. In the presence of catalyst **1c**, an acylative kinetic resolution of a racemic rotaxane **10** afforded a mechanically planar chiral rotaxane **10** with perfect enantiopurity (>99% ee) in 29% yield (the theoretical maximum yield of kinetic resolution of racemate is 50%) (Fig. 3.2). The catalysts enabled to discriminate mobile mechanical chirality of the rotaxanes with the excellent selectivity in up to 16.

Keywords Catalytic synthesis of chiral supramolecules · Rotaxanes and catenanes · Mechanically chiral supramolecules · Mobile supramolecular chirality · Acylative kinetic resolution by organocatalysis · Enantiopure rotaxane

3.1 Backgrounds

A field of asymmetric synthesis has extensively developed during the last few decades. The construction of ubiquitous chiral molecules involving conventional chiral elements such as central, axial or planer chirality has already been adequately realized. With the current state of chemical science, asymmetric synthesis of such fixed chirality over covalent bond is reaching a level of mature science. Chiral supramolecules with additional chiral elements such as central [3, 4], axial [5, 6], planer [7] or helical [8] chirality on their ring or axis component, have already been actively synthesized (Fig. 3.3).

A. Imayoshi, *Discrimination of Mobile Supramolecular Chirality*, Springer Theses,
https://doi.org/10.1007/978-981-16-7431-0_3

Fig. 3.1 Chiral mechanically interlocked molecules

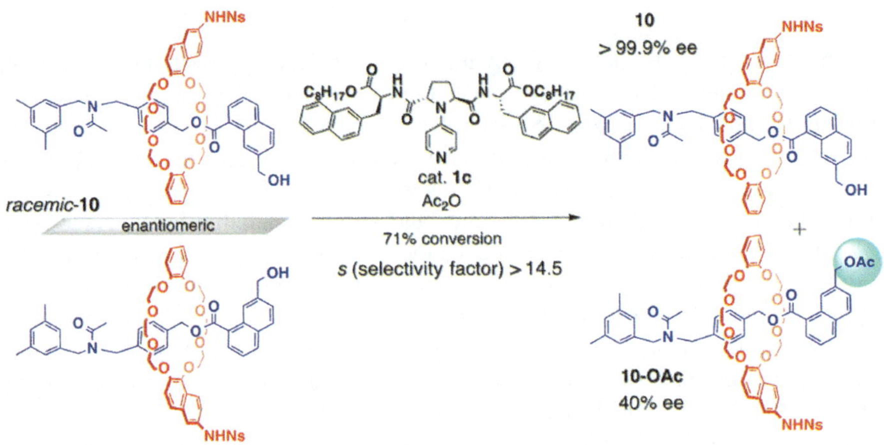

Fig. 3.2 Highly efficient acylative kinetic resolution of racemic mechanically planar chiral rotaxane **10** (The absolute configuration of the acylated product and the recovered rotaxane alcohol were not determined.)

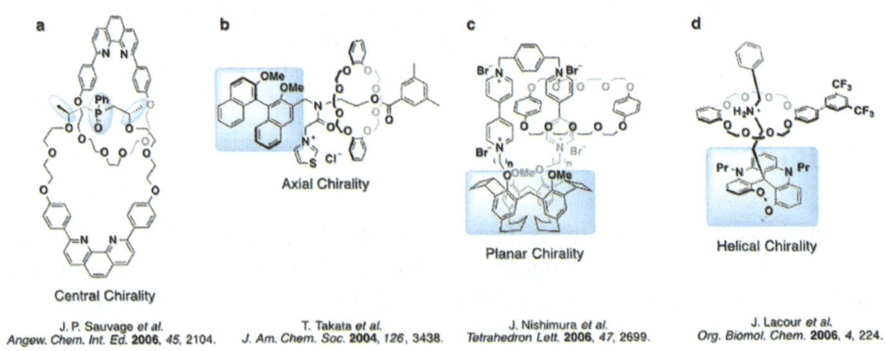

Fig. 3.3 Chiral supramolecules with additional chiral elements. **a** Catenane with central chirality, **b** rotaxane with axial chirality, **c** catenane with planar chirality, **d** rotaxane with helical chirality

On the other hand, even by virtue of current high achievement in the field of asymmetric synthesis, asymmetric construction of chiral supramolecules stand as an unsolved problem.

Mechanically interlocked supramolecules such as rotaxanes and catenanes are known to possess unique supramolecular chirality consisting based on non-covalent binding (Fig. 3.1). *Even though each of the ring and/or axis components has no chiral elements in itself, their combination brings mechanical chirality linked non-covalent system when each of the components has dissymmetry.* Such attractive chiral architectures of supramolecules have captured the attention of synthesis chemists for over six decades. In 1961, catenanes consisting of two dissymmetric ring components were mentioned that they could exist as pairs of enantiomers [1]. These racemic rotaxanes and catenanes were shown to be separable on chiral HPLC [9–11]. However, an efficient asymmetric discrimination and catalytic synthesis of chiral supramolecules still has never been achieved.

Recently, Bordoli and Goldup proposed an excellent practical method for the preparation of an enantiomeric pair of mechanically planar chiral rotaxanes based on diastereomeric synthesis of them. Diastereomeric rotaxanes with mechanically planar chirality and an additional chiral element (i.e., central chirality) were synthesized. Then, the corresponding diastereomers were separated, followed by removal of the chiral auxiliary [12, 13]. The strategy was also applied to the stereoselective synthesis of topologically chiral catenanes [14]. Very recently, Leigh and co-workers achieved enantioselective synthesis of mechanically planar chiral rotaxanes in up to 50% ee by chiral leaving strategy [15]. These diastereomeric strategies require the chiral auxiliary although mechanically chiral supramolecules themselves have no chirality on their ring and/or axis components.

To date, there have been no reports on highly enantioselective catalytic synthesis of mechanically interlocked molecules. The only example of enantioselective catalytic synthesis of rotaxanes has been reported by Takata and co-workers (Fig. 3.4) [16].

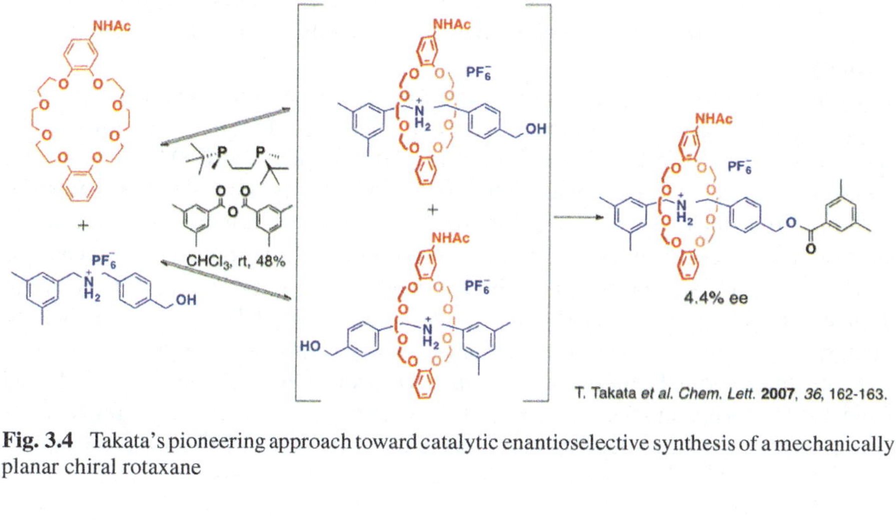

Fig. 3.4 Takata's pioneering approach toward catalytic enantioselective synthesis of a mechanically planar chiral rotaxane

Dynamic kinetic resolution of a racemic pseudorotaxane was performed by catalytic asymmetric acylation of the hydroxy group in the axis component. This pioneering example of discrimination of mechanically planar chirality delivered a rotaxane in up to 4.4% ee.

Thus, enantioselective preparation of mechanically interlocked molecules has been a long-standing challenge in organic synthesis [17]. Under these backgrounds, I made a challenge toward highly enantioselective synthesis of mechanically planar chiral rotaxanes by organocatalysis.

3.2 Mobile Supramolecular Chirality

Asymmetric molecular transformation of mechanically chiral supramolecules is a conventionally difficult subject. The extreme difficulty may be resulting from conformational diversity and movability of mechanically chiral supramolecules. Mechanical chirality is in itself *mobile*, and the kinetically labile mechanically chiral supramolecules continuously change their chiral environments. For example, Fig. 3.5b schematically shows a pair of enantiomers (R_{mp}, S_{mp}) and the chiral conformers (**B**, **E**) of the identical enantiomer (S_{mp}). For asymmetric discrimination of mechanically planar chiral rotaxanes, the relationship of enantiomers (R_{mp}, S_{mp}) should be distinguished, whereas the relationship between the conformers **B** and **E** should *not* be distinguished.

The actual chiral conformers are more complex. The expected typical six chiral conformers **A**–**F** are illustrated in Fig. 3.5c. They all belong to just one single enantiomer (S_{mp})-**10**. Conformers **A**–**F** have different asymmetric microenvironments depending on the location of the ring component. For example, conformers **A** and **D** are pseudo-enantiomeric to each other from the viewpoint of planer chirality. Conformers **E** and **F** are in a similar relationship with **B** and **C**, respectively. A mechanically planar chiral rotaxane has conformational diversity and movability. Thus, it seems to be extremely challenging to discriminate mobile supramolecular chirality.

Such mobile nature of mechanical chirality could be experimentally observed. I measured CD spectra of rotaxane **10-OAc** (Fig. 3.6). Figure 3.6a shows a CD spectra of a pair of separated enantiomers (+)-**10-OAc** and (–)-**10-OAc**, which were measured at 20 °C. A mirror image of the CD spectra clearly showed the enantiomeric relationship of these enantiomers. Figure 3.6b shows the variable-temperature CD spectra of a separated single enantiomer (+)-**10-OAc**. These spectra were measured by heating a MeCN solution of (+)-**10-OAc** from –10 °C to 60 °C. The CD behavior dramatically changed depending on temperature. Molar CD became smaller by raising temperature, although the rotaxane existed as consistently a pure enantiomer (99% ee) throughout the measurement. The CD absorption maximum was red-shifted from 245 to 250 nm and the shape of the CD spectra changed with increasing temperature. Such temperature-dependent CD behavior of rotaxane **10-OAc** was reversible.

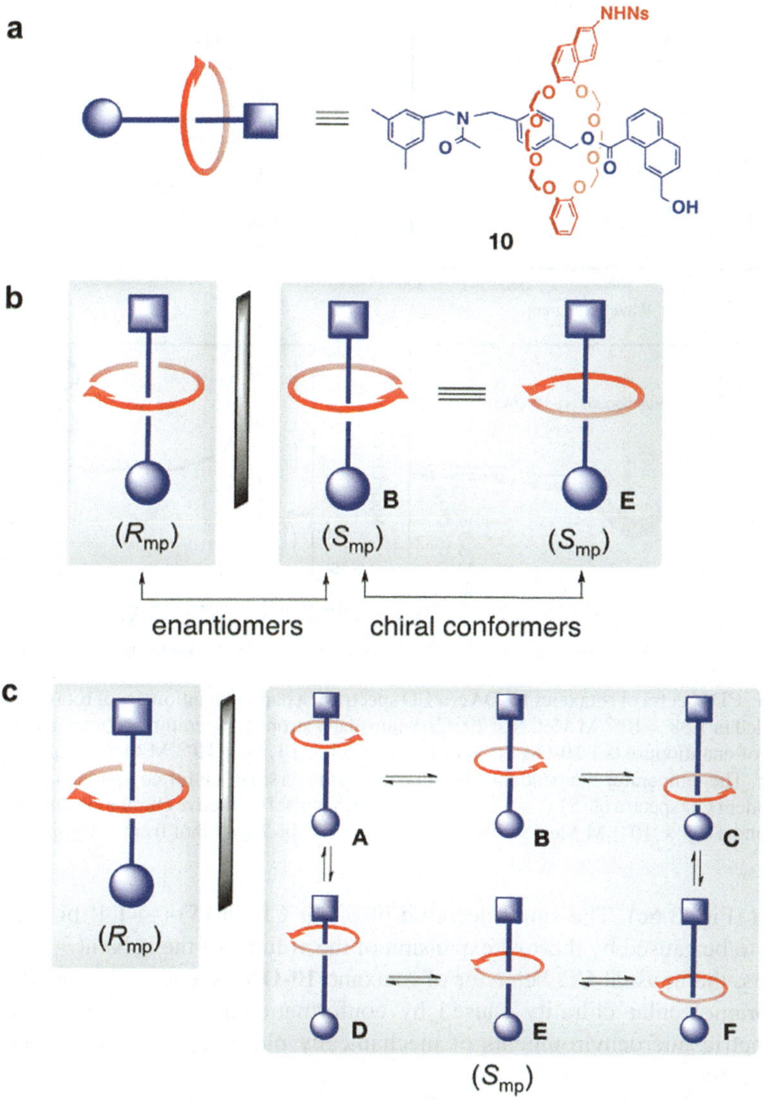

Fig. 3.5 Mobile supramolecular chirality. **a** A schematic presentation of a mechanically planar chiral rotaxane **10**, **b** A pair of enantiomers of rotaxane, and chiral conformers (**B** and **E**) of the identical enantiomer, **c** typical six chiral conformers of a single enantiomer (S_{mp})-**10**

I measured variable-temperature CD spectra of (+)-**10-OAc** by cooling a solution from 60 to −10 °C, and the molar CD increased with decreasing the temperature.

In contrast, (*S*)-(−)-1,1′-bi-2-naphthol, which is known as a typical axially chiral compound, did not show such temperature dependence in a variable-temperature CD

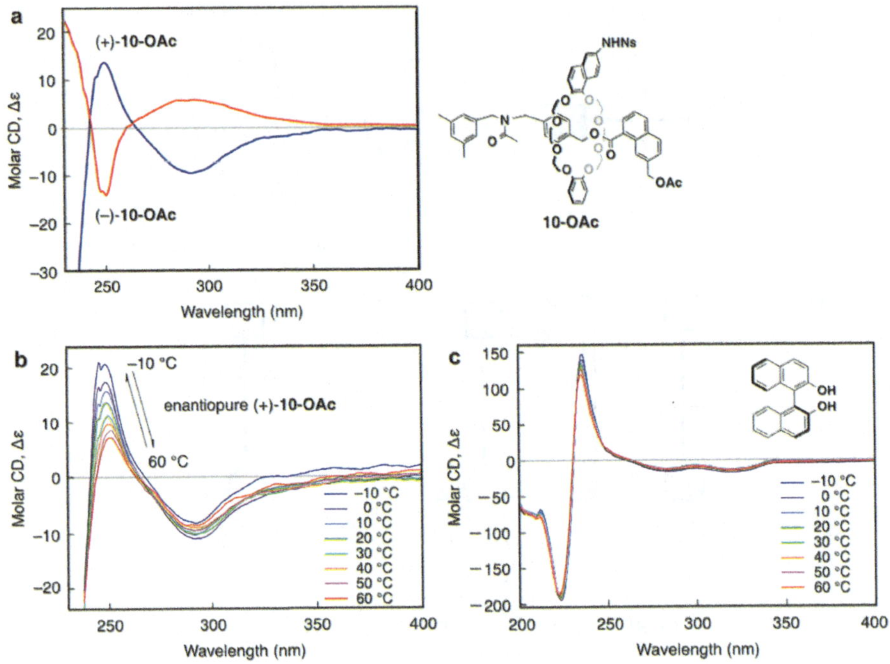

Fig. 3.6 CD spectra of rotaxane **10-OAc**. **a** CD spectra of a pair of enantiomers of rotaxane **10-OAc** (Recorded in 1.38 × 10^{-5} M MeCN at 20 °C), **b** unusually strong temperature-dependency of the CD spectra of enantiopure (+)-**10-OAc** (Recorded by heating a 1.38 × 10^{-5} M MeCN solution of (+)-**10-OAc**. The temperature-dependent CD behavior was reversible.), **c** for comparison, temperature independent CD spectra of (*S*)-(–)-1,1′-bi-2-naphthol, a typical CD active chiral molecule (Recorded by heating a 0.5 × 10^{-5} M MeCN solution of (*S*)-(–)-1,1′-bi-2-naphthol from −10 to 60 °C)

spectra (Fig. 3.6c). The small decrease of molar CD of (*S*)-(–)-1,1′-bi-2-naphthol seems to be caused by thermal expansion of the volume of the solvent.

Thus, the unusual CD behavior of rotaxane **10-OAc** suggests the mobile nature of supramolecular chirality caused by conformational diversity and changes in asymmetric microenvironments of mechanically planar chiral rotaxane as shown in Fig. 3.5c.

Based on these observation of CD spectra, I measured a variable-temperature NMR of rotaxane **10-OAc** in acetonitrile-d_3 in the temperature range from −40 to 60 °C (Figs. 3.7, 3.8 and 3.9). The assignments of the ^{1}H NMR spectra of rotaxane **10-OAc** were made using COSY, HMQC, HMBC and DEPT 135 methods at 20 °C. The ^{1}H NMR spectra of **10-OAc** showed strong temperature dependency. Upon heating, many pairs of signals coalesced or broadened. Notably, these chemical shifts at the coalescence temperature were not the average chemical shifts of the corresponding individual two signals observed at the low temperature. For example, benzylic proton signals Hb (Fig. 3.8), which were observed at ca. 6.24 and 5.92 ppm at −40 °C, appeared at 6.21 and 6.18 ppm at 60 °C. The average chemical shift (6.08 ppm) at –

Fig. 3.7 Variable-temperature NMR spectra (600 MHz) of **10-OAc** in CD_3CN from −40 °C to 60 °C (10–6.5 ppm)

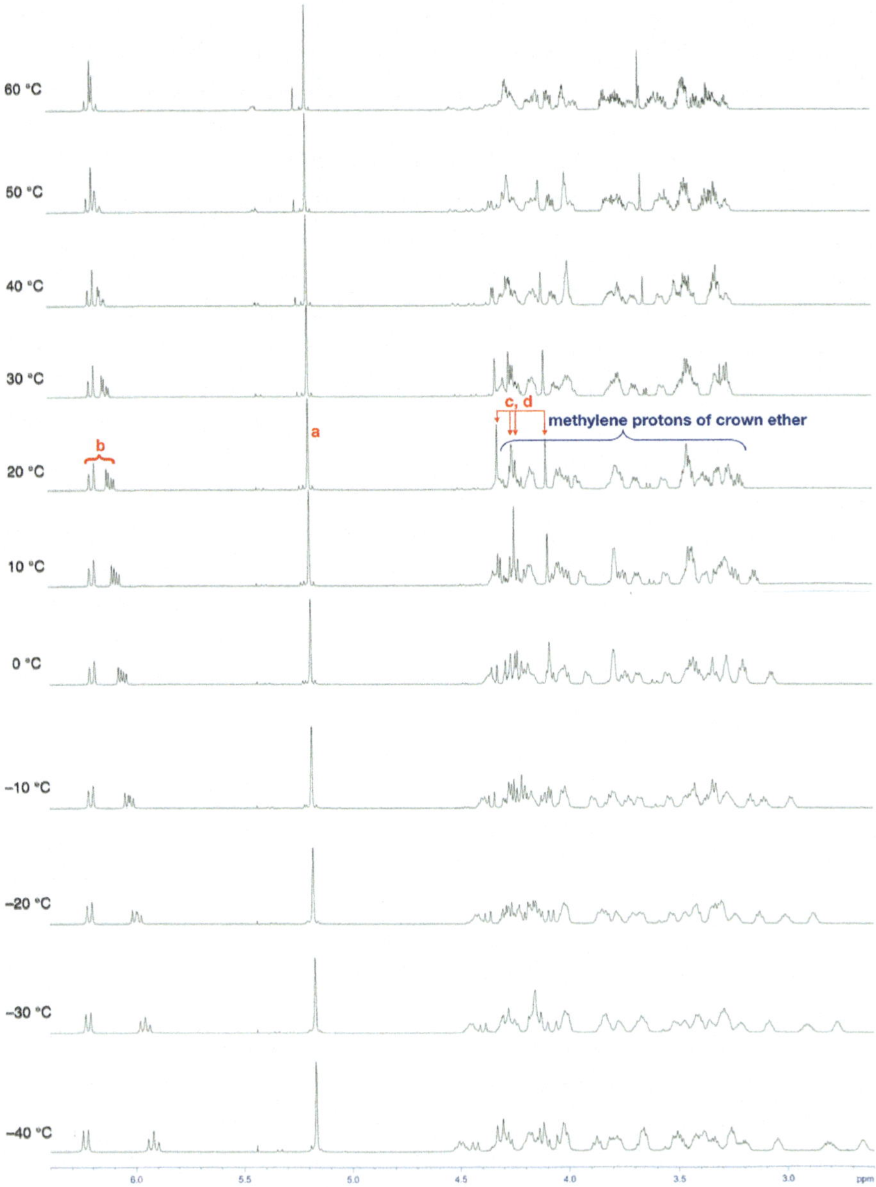

Fig. 3.8 Variable-temperature NMR spectra (600 MHz) of **10-OAc** in CD_3CN from −40 °C to 60 °C (6.5–2.5 ppm)

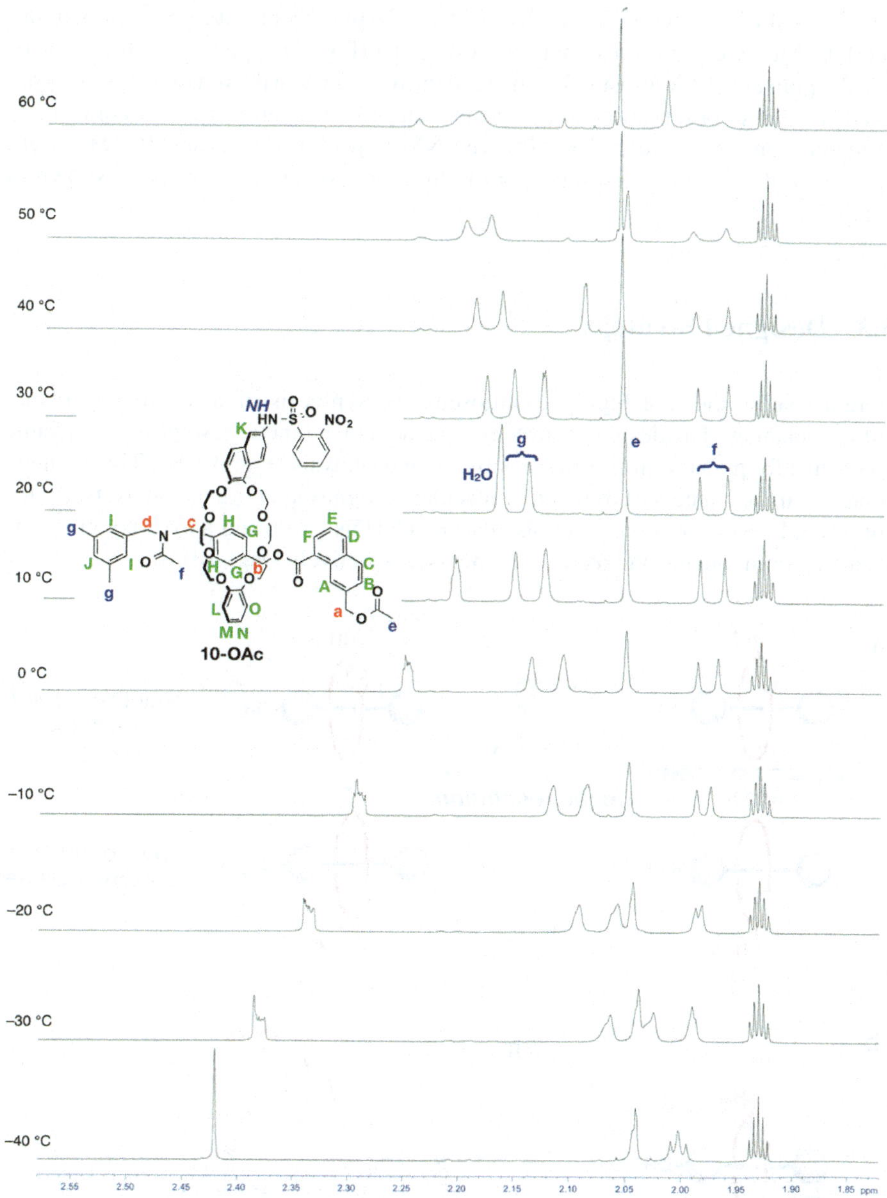

Fig. 3.9 Variable-temperature NMR spectra (600 MHz) of **10-OAc** in CD_3CN from −40 °C to 60 °C (2.5–1.9 ppm)

40 °C was shifted downfield and observed at 6.20 ppm by elevating the temperature at 60 °C. Methylene proton signals of crown ether (Fig. 3.8) appeared in the range of 4.6–2.6 ppm at −40 °C drastically changed the resonance in the range of 4.3–3.3 ppm at 60 °C. These observations suggest the change of the ratio of chiral conformers depending on temperature. The observed NMR spectra of rotaxane **10-OAc** seem to indicate the diversity and mobility of chiral environment of rotaxanes shown in Fig. 3.5c.

3.3 Designed Strategy

Herein I show the first highly enantioselective synthesis of mechanically planar chiral rotaxanes. The designed strategy is an acylative kinetic resolution of racemic mechanically planar chiral rotaxanes by organocatalysis (Fig. 3.10a). This strategy is based on the concept of dynamic molecular recognition of organocatalysts. As the substrate, I chose rotaxane alcohols bearing a NHNs group and a hydroxy group on their ring components and axis components, respectively (Fig. 3.10b).

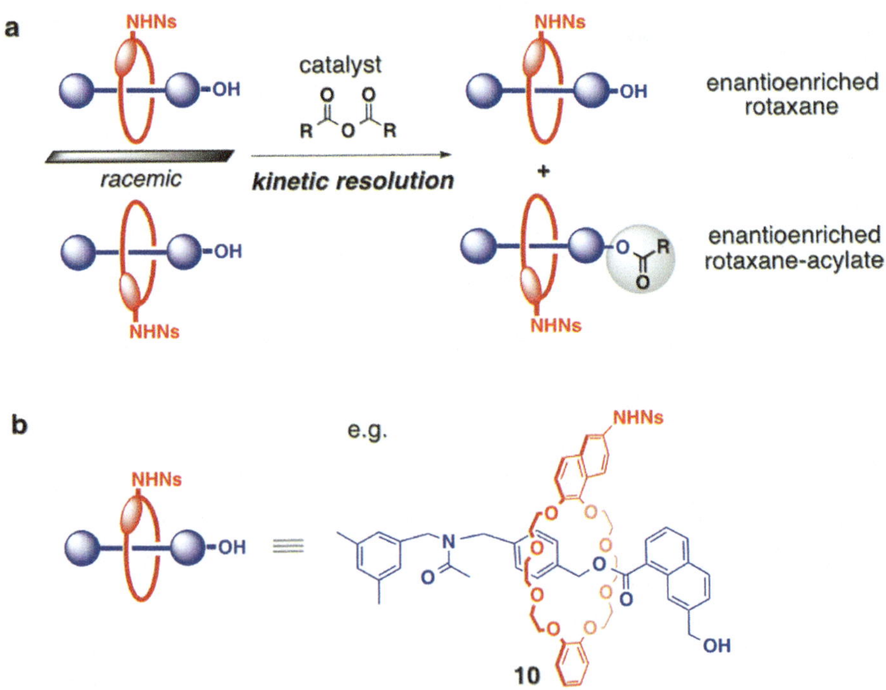

Fig. 3.10 Designed strategy. **a** Kinetic resolution of a mechanically chiral racemic rotaxane by organocatalysis, **b** example of rotaxane substrates for kinetic resolution

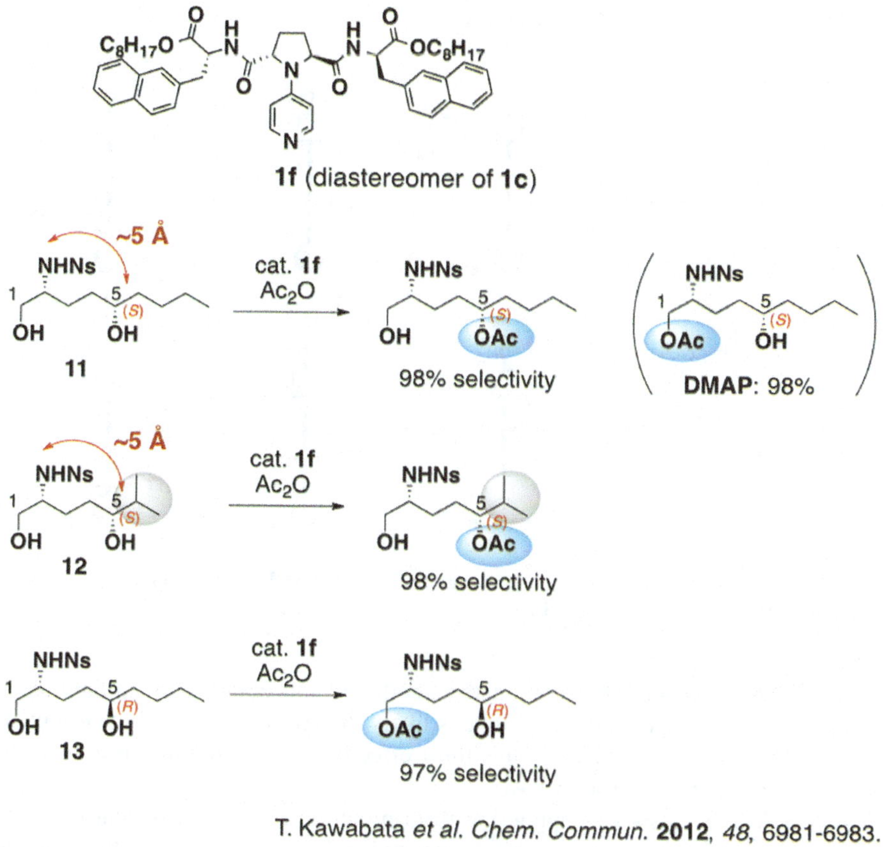

Fig. 3.11 A hint for dynamic molecular recognition from previous work in our laboratory

As the backgrounds, our laboratory previously reported reversal of chemoselectivity in the acylation of diols with an NHNs group (Fig. 3.11) [18]. Acylation of diol **11** catalyzed by 4-dimethylaminopyridine (DMAP) took place at the less hindered primary hydroxy group with 98% chemoselectivity; i.e., normal chemoselectivity. On the other hand, highly chemoselective (98%) acylation of a more hindered secondary alcohol of **11** proceeded in the presence of catalyst **1f**. Further reversal of the chemoselectivity of acylation was observed in sterically more demanding alcohol **12**. These results and other results regarding chemoselective acylation observed in our laboratory [18–21] suggest that catalyst **1f** and its analogues can promote chemoselective acylation of a hydroxy group located at a certain distance (~5 Å) from the nitrogen atom of the NHNs group.

In addition, catalyst **1f** and its analogues seem to be capable of recognizing remote chirality at the reacting center (hydroxy group). For example, highly chemoselective acylation of the secondary hydroxy group of diols (5*S*)-**11** and (5*S*)-**12** took place in the presence of catalyst **1f**, while acylation of (5*R*)-**13** occurred at the

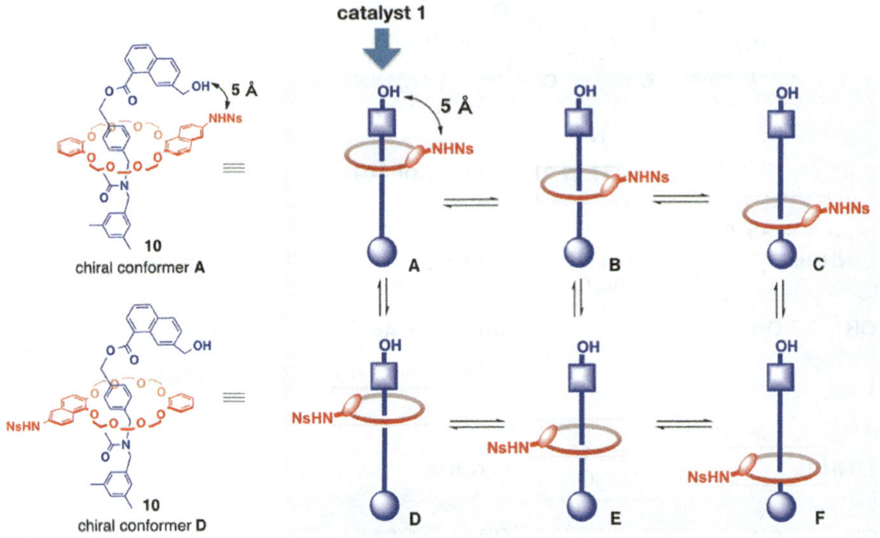

Fig. 3.12 Concept of dynamic molecular recognition by catalyst **1** for discrimination of mobile supramolecular chirality of rotaxane **10**

primary hydroxy group [18]. These data and other results obtained in our laboratory suggest that catalyst **1f** and its analogues likely discirmininate chiral environment consisting of the OH and NH groups most effectively when the distance between theses functionalities is around 5 Å.

This 5 Å hypothesis was applied to discrimination of supramolecular chirality (Fig. 3.12). Rotaxanes have mobile supramolecular chirality, thus, the relative topology between the NHNs group and the hydroxy group continuously changes. However, at the moment when these functionalities reach the distance of ~5 Å, the catalyst would be able to discriminate the chiral environment of the rotaxane.

Based on this concept of dynamic molecular recognition, I envisaged a kinetic resolution of racemic mechanically planar chiral rotaxanes by organocatalysis.

3.4 Synthesis of Substrates

First of all, I synthesized rotaxane substrates bearing a hydroxy group on their axis components and a NHNs group on their ring components. Rotaxanes **10**, **14** and **15** were synthesized from the corresponding ring components **20–22**, axis component **19**, and endcap **23** (Fig. 3.13).

Ring components were prepared from 6-bromo-2-naphthol (Fig. 3.14). Oxidation of 6-bromo-2-naphthol gave the diketone **24**, which was reduced into naphthalene-1,2-diol **25**. Since naphthalene-1,2-diol **25** was very sensitive to air, **25** was carefully treated under Ar atmosphere. Naphthalene-1,2-diol **25** was converted to stable

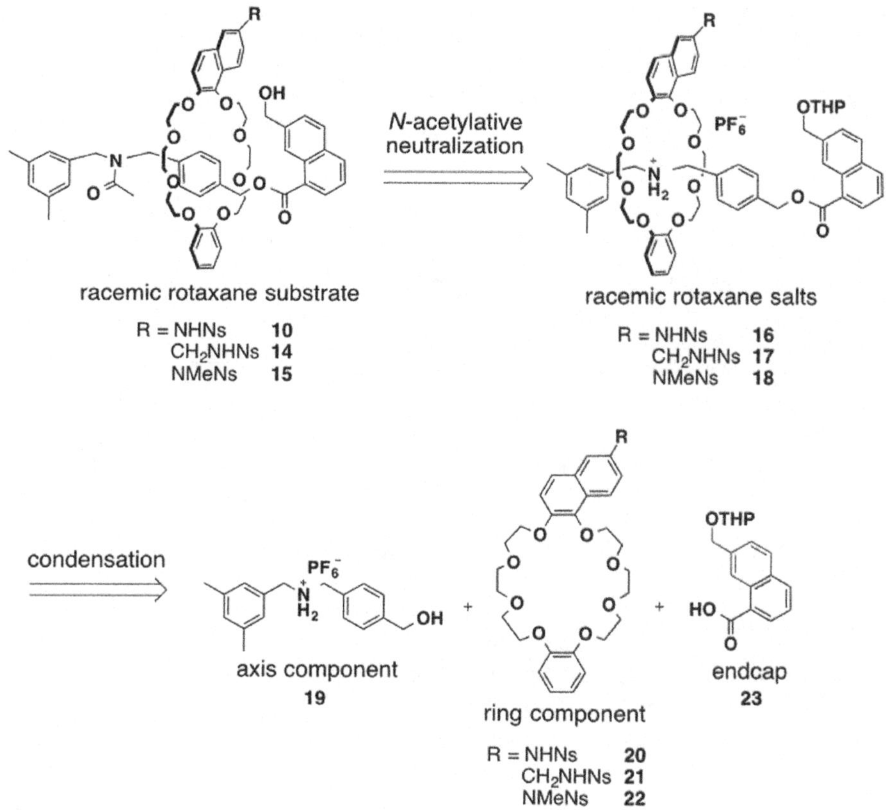

Fig. 3.13 Retrosynthesis of rotaxane substrates **10**, **14**, and **15**

alcohol **26**. Conversion of alcohol **26** to crown ether **28** was performed by tosylation of the hydroxy groups, followed by cesium cation-templated cyclization. The cyclization reaction required at least one week for the completion. Buchwald-Hartwig amination of naphthylbromide **28**, followed by deprotection of the Boc group of **29**, provided amine hydrochloride **30**. Treatment of amine **30** with *o*-NsCl gave desired ring component **20** (9.7 g).

Ring component **22** bearing *N*-methyl substituent was prepared from the ring component **20** (Fig. 3.14a).

Ring component **21** was prepared from naphthylbromide **28** in three steps (Fig. 3.14b). Rosenmund-von Braun reaction of naphthylbromide **28**, followed by reduction of nitrile **31**, provided naphthylmethylamine **32**. Naphthylmethylamine **32** was converted into desired ring component **21** bearing *o*-Ns substituted naphthylmethylamine.

THP-protected endcap **23** was prepared from 1,7-dihydroxynaphthalene (Fig. 3.15). Triflation of 1,7-dihydroxynaphthalene followed by carbonylation gave dimethyl ester **34**. Monosaponification of diester **34** gave desired carbonyl compound

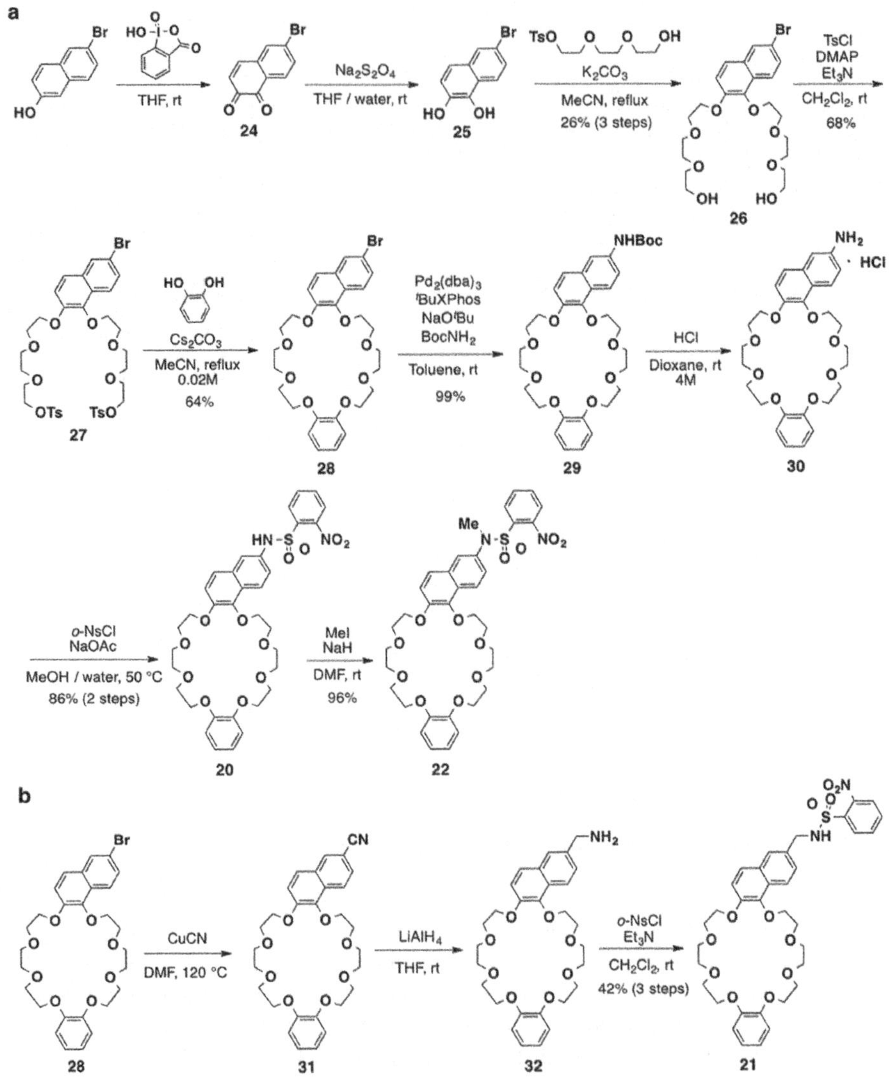

Fig. 3.14 Synthesis of ring components **a 20**, **22**, and **b 21**

35 after reprecipitation. Reduction to alcohol **36** and THP protection of the hydroxy group of **36** preceded the saponification of methyl ester **37**, which led to the desired endcap **23** (5.0 g).

Reported axis component **19** was prepared from 3,5-dimethylbenzoic acid (Fig. 3.16) [22]. 3,5-Dimethylbenzoic acid was converted to amide **39** followed by reduction to give benzylamine **40**. Conversion of benzylamine **40** to amino alcohol **42** was performed by reductive amination by using lithium aluminum hydride, where the methyl ester of **41** was also reduced to the corresponding alcohol. Amino alcohol

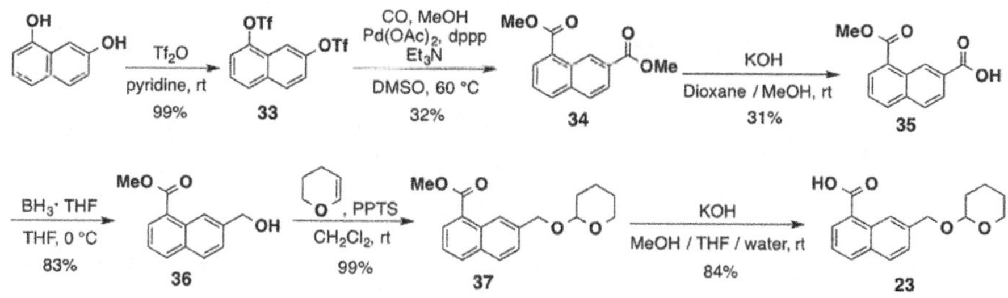

Fig. 3.15 Synthesis of endcap component **23**

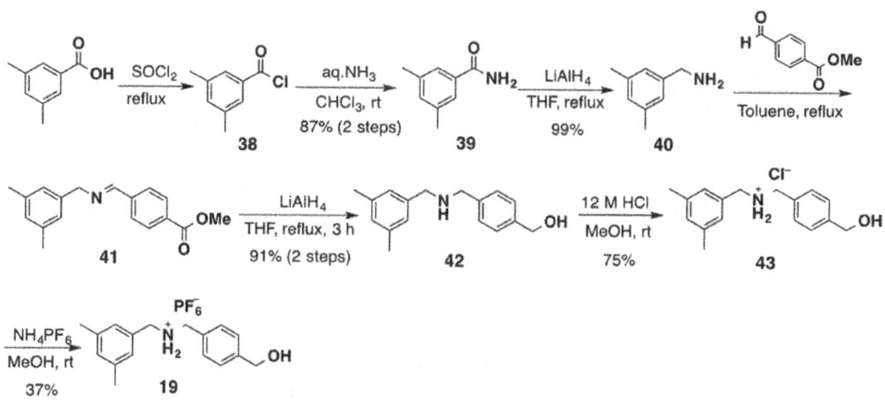

Fig. 3.16 Synthesis of axis component **19**

42 was converted into desired axis component **19** via corresponding hydrochloride salt **43**.

Then, racemic rotaxane **10** was synthesized from the corresponding ring component **20**, axis component **19** and endcap **23** (Fig. 3.17a). Rotaxane **16** was prepared according to the tributylphosphine-catalyzed acylative end-capping protocol reported by Takata and co-workers [22]. In this step, rotaxane **16** becomes a racemic mixture because the axis component threads the ring component randomly from both sides of the ring component. Racemic rotaxane **16** was converted to **44** by *N*-acylation of the axis component. Since the amino group on the ring component was also acylated as well as the amino group on the axis component, selective hydrolysis of the amide group on the ring component was performed to give **46**. Finally, deprotection of THP group of **46** provided the desired racemic rotaxane substrate **10** (1.5 g).

Rotaxane substrate **14** (Fig. 3.17a) and **15** (3.17b) could be prepared by the similar manner employed for **10**.

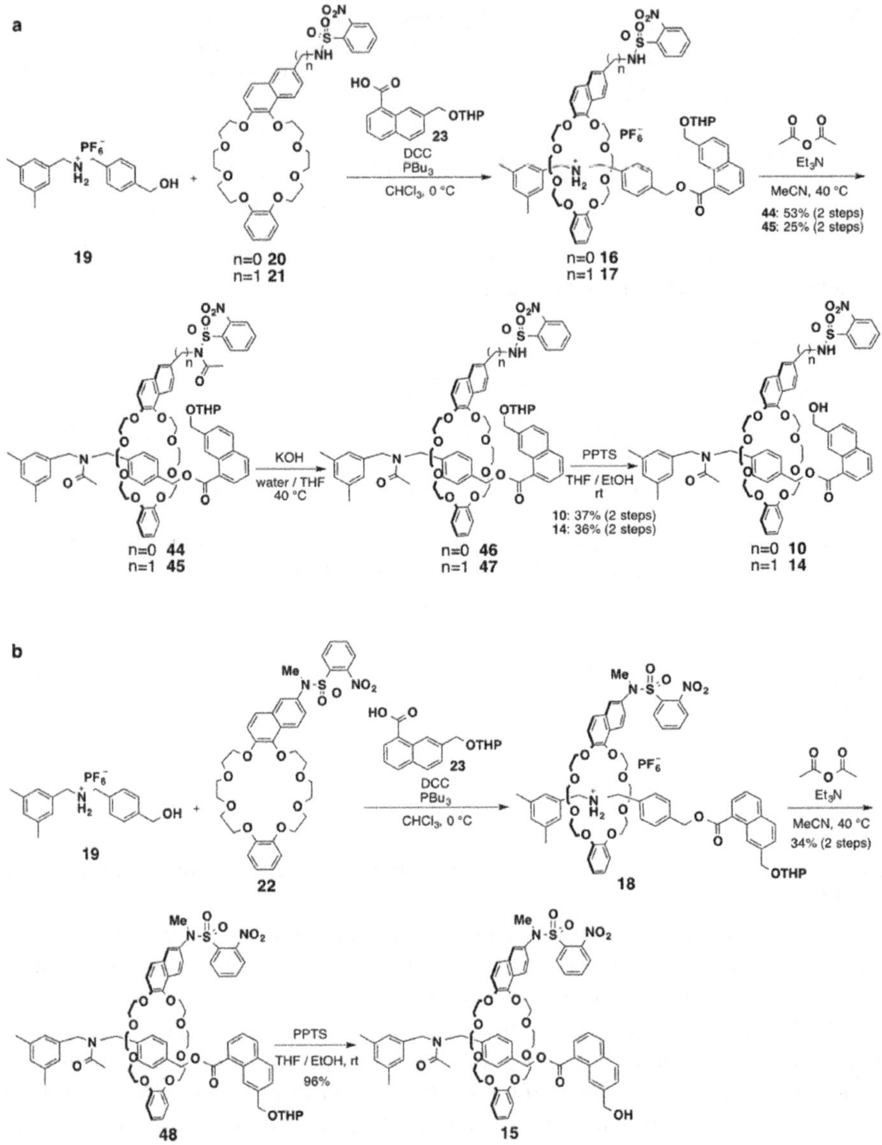

Fig. 3.17 Synthesis of *racemic* rotaxanes **a 10**, **14**, and **b 15**

3.5 Optimization of Conditions

3.5.1 Screening of Catalysts

I initially examined various 4-pyrroridinopyridine (PPY)-based catalysts **1c**, *ent*-**1a**, and **1f–1 h** for acylative kinetic resolution of racemic mechanically planar chiral rotaxane **10** (Fig. 3.18). Among them, catalyst **1c** was found to be the most effective catalyst, which gave acylated **10-OAc** and recovered-**10** with 77% ee and 15% ee, respectively, with selectivity factor of 2.6 (entry 1). Catalyst **1f**, the diastereomer of catalyst **1c**, was found less effective for the kinetic resolution (s = 1.4). Catalysts

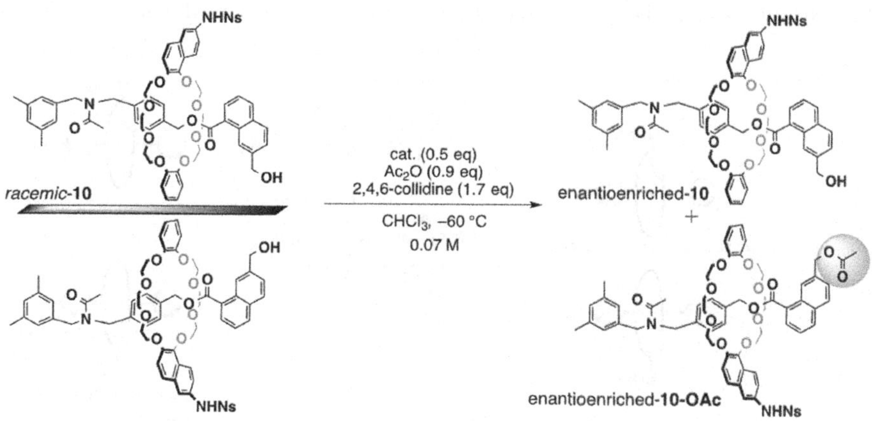

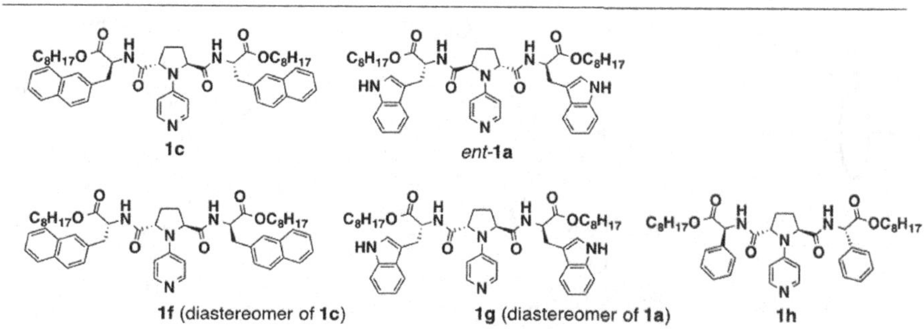

entry	cat.**1**	ee of **10-OAc** (%)	ee of **10** (%)	conversion (%)	*s* factor
1	**1c**	15	77	83	2.6
2	*ent*-**1a**	4	22	84	1.3
3	**1f**	9	19	67	1.4
4*	**1g**	−7	−39	85	1.5
5	**1h**	21	12	36	1.7

* Run in 0.007 M solution.

Fig. 3.18 Screening of catalysts

ent-**1a**, **1 g** and **1 h**, which are replaced the naphthyl side chain of catalyst **1c** with other substituents, also showed diminished enantioselectivity ($s = 1.3$–1.7).

3.5.2 Screening of Substrates

I next examined the substrate structure. Rotaxane alcohols **10**, **49** and **50** possess an amide group with various degree of steric bulkiness on their axis components (Fig. 3.19). I expected that the bulky amide group (R = Ph, 1-adamantyl) on the axis component improves enantioselectivity by pushing the ring component to the edge of the axis component, thereby the NHNs group easily approaches the hydroxy group at the distance of 5 Å. However, steric bulkiness of amide groups did not provide any

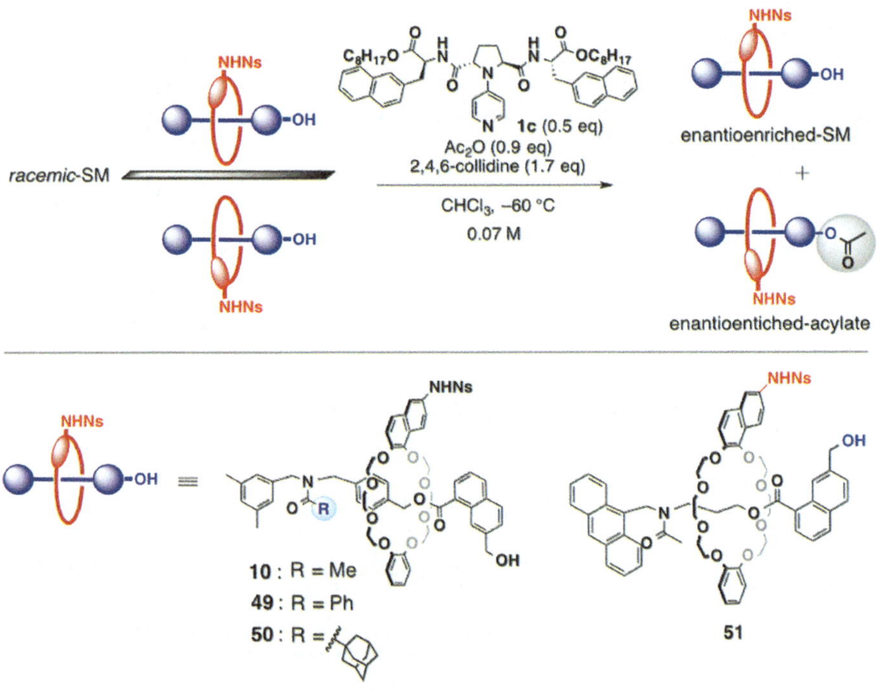

entry	Substrate	ee of acylate (%)	ee of SM (%)	conversion (%)	*s* factor
1	**10**	15	77	83	2.6
2	**49**	10	43	81	1.7
3	**50**	8	65	89	1.9
4*	**51**	17	18	51	1.7

* Cat. **1h** was used instead of cat. **1c**. Run in 0.007 M solution.

Fig. 3.19 Initial screening of structure of substrates

racemic-SM —[cat. **1c**, acylating agent, base / solvent, −60 °C]→ enantioenriched-SM + enantioenriched-acylate

Entry	rotaxane substrate	concentration (M)	cat. **1c** (equiv)	acylating agents (equiv)	base (equiv)	solvent	*s*
1	Me (**10**)	0.01	0.5	Ac_2O (0.5)	2,4,6-collidine (1.2)	$CHCl_3$	6.1
2	Me (**10**)	0.01	1.5	Ac_2O (0.5)	triethylamine (1.2)	$CHCl_3$	–*
3	Me (**10**)	0.07	0.5	Ac_2O (0.9)	2,4,6-collidine (1.7)	$CHCl_3$	2.6
4	Ad (**50**)	0.07	0.5	Ac_2O (0.9)	2,4,6-collidine (1.7)	$CHCl_3$	1.9
5	Ad (**50**)	0.07	0.5	$(^{i}PrCO)_2O$ (0.9)	2,4,6-collidine (1.7)	$CHCl_3$	1.8
6	Me (**10**)	0.005	0.5	Ac_2O (0.5)	2,4,6-collidine (1.7)	toluene/$CHCl_3$ (20/1)	1.6
7	Me (**10**)	0.005	0.5	Ac_2O (0.5)	2,4,6-collidine (1.7)	CS_2/$CHCl_3$ (20/1)	1.2
8	Me (**10**)	0.006	1.5	Ac_2O (0.5)	2,4,6-collidine (1.7)	$CHCl_3$	6.3
9	Me (**10**)	0.006	1.5	Ac_2O (0.5)	2,4,6-collidine (1.2)	$CHCl_3$	6.6
10	Me (**10**)	0.006	1.5	Ac_2O (0.5)	2,4,6-collidine (0.9)	$CHCl_3$	6.3

* By-products were observed, and the selectivity factor (*s*) was low.

Fig. 3.20 Investigation of reaction parameters

positive effects on efficiency of kinetic resolution ($s = 1.7, 1.9$). Rotaxane substrate **51** also showed no significant improvement ($s = 1.7$).

3.5.3 Investigation of Reaction Parameters

Having revealed that catalyst **1c** is effective in discrimination of mobile supramolecular chirality, I then investigated variation of reaction parameters, including solvent, acylating agent, base, and concentration (Fig. 3.20). The optimum conditions were found to be the treatment of **10** with acetic anhydride and 1.2 equivalents of 2,4,6-collidine in the presence of catalyst **1c** in 0.01 M $CHCl_3$ at −60 °C.

3.5.4 Effects of Catalyst Loading

In an effort to obtain higher enantioselectivity, a breakthrough in optimization of conditions came with evaluation of catalyst loading. I initially conducted the kinetic resolution using 0.05 equivalents of catalyst **1c** for 6 h, which afforded selectivity factor of 2.2, but the conversion was very low (3%, Fig. 3.21, entry 2). For the purpose to increase the reaction conversion, the catalyst loading was then increased to 0.2 equivalents. The conversion was increased to 9% as expected, but the selectivity factor was also unexpectedly improved to 4.2 (entry 3). In usual kinetic resolutions, selectivity factors do not change depending on the catalyst loading unless the reaction proceeds in the absence of catalysts, thus this is an unusual tendency (entry 1) [23, 24].

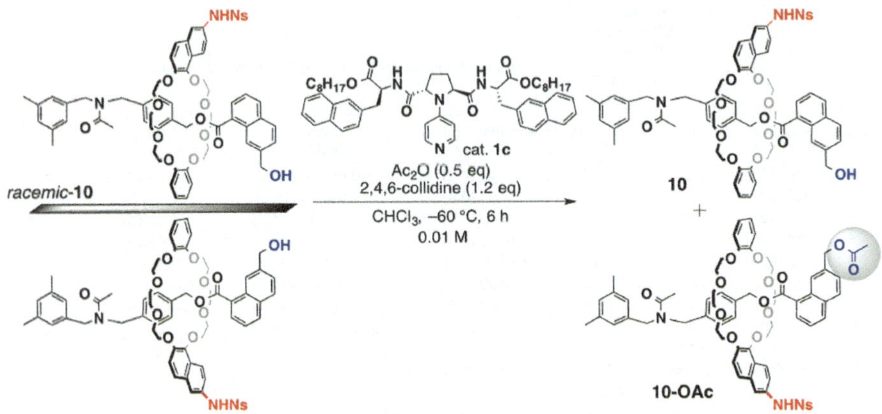

entry	cat.**1c** (eq.)	ee of **10-OAc** (%)	ee of **10** (%)	conversion (%)	*s* factor
1	none	-	-	<1	-
2	0.05	38	1	3	2.2
3	0.2	60	5	9	4.2
4	0.5	66	23	26	6.1
5	1.5	79	63	44	16.1
6	2.0	63	58	48	7.8

Fig. 3.21 Effects of catalyst loading

Based on this curious observation, I investigated the effects of the catalyst loading in detail. Finally, 1.5 equivalents of catalyst **1c** was found to be most effective in asymmetric discrimination of mechanically planar chiral rotaxane **10**, which afforded the **10-OAc** in 79% ee and recovered **10** in 63% ee at 44% conversion with an outstanding level of selectivity factor of 16.1 (entry 5). I consider that the kinetic resolution likely proceeded by the mechanism in which more than two molecules of catalyst **1c** are involved. Although 1.5 equivalents of catalyst **1c** were used, only a small amount of catalytically active species (acylpyridinium salts) seemed to be produced in the reaction system (see Appendix: Chap. 4).

3.5.5 *Effects of Conversion*

In the course of reactions, the degree of conversions did not affect the selectivity factor (Fig. 3.22). Catalyst **1c** kept the high level of selectivity factor ($s = 16$) regardless of the degree of conversions. Thus, the enhanced enantioselectivity shown in Fig. 3.21 seems to result from the effects of the catalyst loading, not from the degree of conversions.

racemic-**10** $\xrightarrow[\text{CHCl}_3,\ -60\ ^\circ\text{C},\ 0.01\ \text{M}]{\text{cat. } \mathbf{1c}\ (1.5\ \text{eq}),\ \text{Ac}_2\text{O}\ (0.5\ \text{eq}),\ 2,4,6\text{-collidine}\ (1.2\ \text{eq})}$ enantioenriched-**10** + enantioenriched-**10-OAc**

time (h)	ee of **10-OAc** (%)	ee of **10** (%)	conversion (%)	*s* factor
2	85	29	25	16.3
4	83	38	31	15.6
6	79	63	44	16.1

Fig. 3.22 Effects of conversion on selectivity factors (*s*)

3.5.6 Effects of Concentration

The concentration of the reaction mixture also significantly affected the enantioselectivity as well as the catalyst loading (Fig. 3.23).

The optimum substrate concentration of the reaction was 0.01 M (entry 2), where the catalyst loading greatly affected the selectivity factors (Fig. 3.21). However, kinetic resolution at lower concentration (0.001 M) resulted in the reduced enantioselectivity ($s = 4.4$) despite the use of 1.5 equivalents of catalyst **1c** (Fig. 3.23, entry 1). I suspect that the loss in enantioselectivity is due to the predominance of the unimolecular catalytic process rather than bi-molecular catalytic process at the lower concentration. Kinetic resolution at higher concentration (0.1 M) also provided a reduced enantioselectivity ($s = 2.3$–2.6) regardless of the catalyst loading (entries 3, 4). The loss in the enantioselectivity is probably due to the dimeric association of the substrate at the high concentration, which may not be suitable for proper substrate–catalyst interaction. Thus, the catalyst loading was found to be critical only at the optimized concentration (0.01 M).

racemic-**10** $\xrightarrow[\text{CHCl}_3,\ -60\ ^\circ\text{C},\ 6\ \text{h}]{\text{cat. } \mathbf{1c},\ \text{Ac}_2\text{O}\ (0.5\ \text{eq}),\ 2,4,6\text{-collidine}\ (1.2\ \text{eq})}$ enantioenriched-**10** + enantioenriched-**10-OAc**

entry	**10** (M)	cat. **1c** (M)	cat. **1c** (eq.)	ee of **10-OAc** (%)	ee of **10** (%)	conversion (%)	*s* factor
1	0.001	0.0015	1.5	62	0.3	5	4.4
2	0.01	0.015	1.5	79	63	44	16.1
3	0.1	0.15	1.5	34	25	42	2.6
4	0.1	0.015	0.15	32	18	36	2.3

Fig. 3.23 Effects of concentration of rotaxane substrate **10** and catalyst **1c**

3.6 Preparative Scale Production of an Enantiopure Rotaxane

With such a high selectivity in the kinetic resolution of the rotaxane with mobile supramolecular chirality in hand, it is now possible to obtain mechanically planar chiral rotaxane **10** in excellent ee with good yield. To demonstrate the utility of the present method, I performed the reaction on preparative-scale (Fig. 3.24). 50 mg of racemic mechanically planar chiral rotaxane **10** was used for kinetic resolution. The reaction was performed with 1.5 equivalents of **1c** and 0.8 equivalents of acetic anhydride in the presence of 1.2 equivalents of 2,4,6-collidine in $CHCl_3$ at −60 °C for 37 h. As a result, 15 mg of mechanically planar chiral rotaxane **10** was successfully isolated with perfect enantiopurity (>99.9% ee) in 29% yield (the theoretical maximum yield of kinetic resolution of racemate is 50%) with a selectivity factor of >14.5.

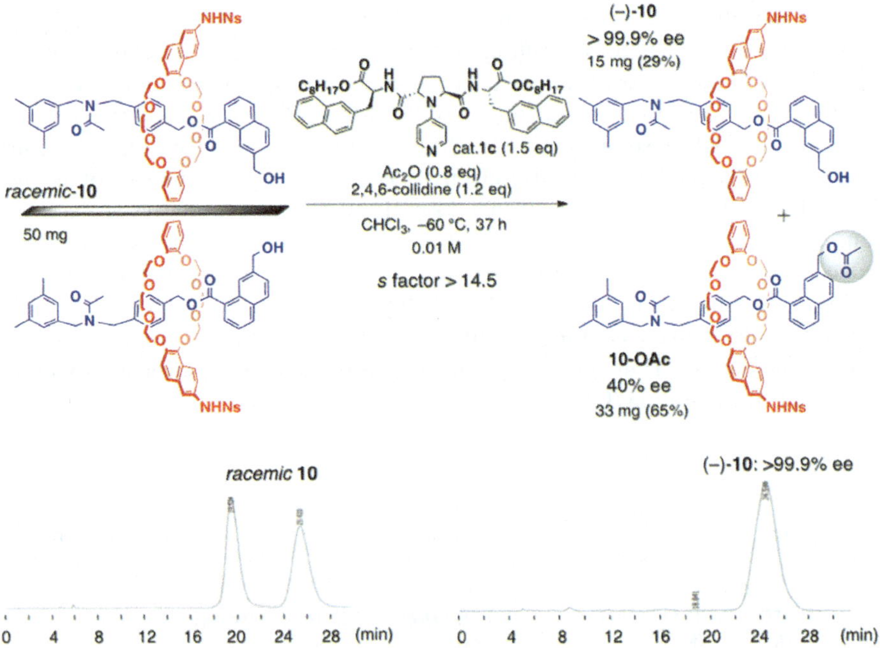

Fig. 3.24 Preparative scale production of enantiomerically pure rotaxane **10** and the chiral HPLC chromatograms (The absolute configuration of the acylated product and the recovered rotaxane alcohol were not determined. Conditions for HPLC: CHIRALPAK IC column (4.6 × 250 mm); eluent 1:60 $EtOH/CH_2Cl_2$; flow rate 0.7 ml/min; detection 254 nm, temperature 20 °C)

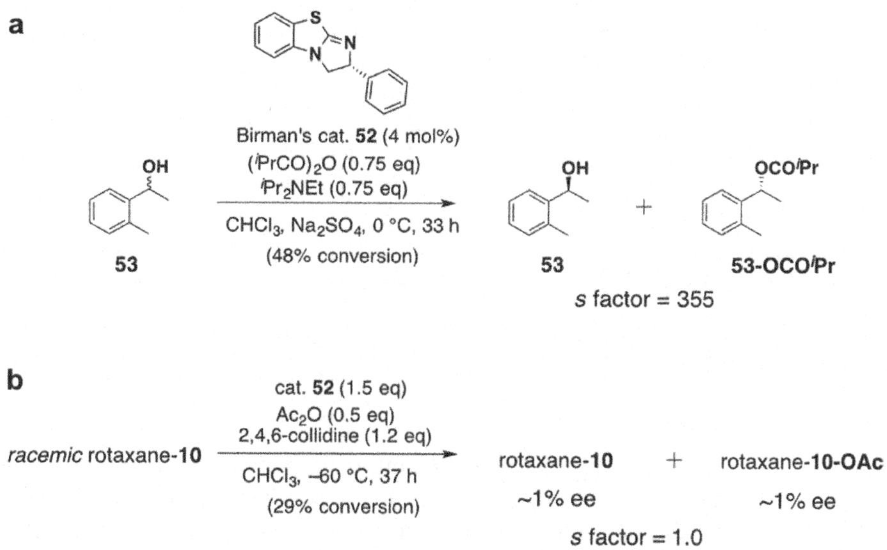

Fig. 3.25 Acylative kinetic resolution of alcohols with Birman's catalyst **52**. **a** Highly selective acylative kinetic resolution of alcohol **53** with Birman's catalyst **52**, **b** kinetic resolution of mechanically chiral racemic rotaxane **10** with Birman's catalyst **52**

3.7 Kinetic Resolution of a Rotaxane with a Typical Catalyst

I also performed kinetic resolution of rotaxane **10** in the presence of Birman's catalyst **52** instead of catalyst **1c**. Acylative kinetic resolution of racemic alcohols has been extensively developed, and Birman's catalyst has been well recognized as one of the most efficient catalysts for the purpose. Actually selectivity factor reaching 355 has been reported for the acylative kinetic resolution of a racemic alcohol **53** with Birman's catalyst **52** (Fig. 3.25a) [25]. However, catalyst **52** was totally ineffective for the kinetic resolution of rotaxane **10**, which afforded **10-OAc** and recovered **10** as the racemate with selectivity factor of 1.0 (Fig. 3.25b).

3.8 Effects of Acidic Hydrogens of NHNs Groups

In order to gain some mechanistic insights, I investigated the effects of the NHNs group of **10** on discrimination of mobile supramolecular chirality by catalyst **1c** (Fig. 3.26). Kinetic resolution of rotaxane **15**, bearing the NMeNs group instead of the NHNs group, was performed under the optimum conditions. As the result, much less selectivity ($s = 1.3$) than that for **10** ($s = 16.1$) was observed. The hydrogen of the NHNs group in **10** appears to play an important role for achieving highly efficient

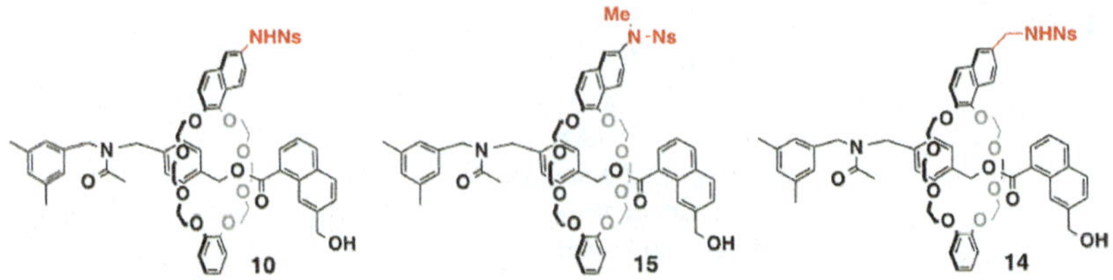

	10	15	14
s factor	16.1	1.3	3.6
conversion (time)	44% (6 h)	8% (37 h)	48% (6 h)
(acylate, SM)	79% ee, 63% ee	11% ee, 1% ee	43% ee, 39% ee

Fig. 3.26 Effects of acidic hydrogens of NHNs groups (Reaction conditions: catalyst **3** (1.5 equivalents), acetic anhydride (0.5 equivalents), 2,4,6-collidine (1.2 equivalents), $CHCl_3$, −60 °C, 0.01 M)

discrimination of mobile supramolecular chirality. Formation of a hydrogen bond between the NHNs group and catalyst **1c** may be the possible origin of the observed phenomena. In the kinetic resolution of **15**, conversion was also reduced (8%, 37 h) as well as the enantioselectivity compared to that of **10** (44%, 6 h). This result indicates that the NHNs group in **10** is responsible for the accelerative acylation.

I also performed kinetic resolution of **14** with the benzylic NHNs group, and the modest enantioselectivity was observed ($s = 3.6$). The more acidic hydrogen of the NHNs group in **10** than that in **14** seems to work as a more efficient hydrogen bond donor for the interaction with catalyst **1c**.

3.9 A Proposal of a Model for Chiral Discrimination

Catalyst **1c** worked as a highly effective catalyst for discrimination of mechanically planar chirality of rotaxane **10**. In contrast, catalyst **52** was totally ineffective in the kinetic resolution although Birman's catalyst **52** has been known to be one of the best catalysts for acylative kinetic resolution of alcohols. Difference in the modes for the chiral discrimination between Figs. 3.25a and 3.24 is shown in Fig. 3.27a. On the case of chiral discrimination of alcohol **53**, the chiral center is located immediately adjacent to the reacting center, so that high performance in the discrimination of chirality of the molecule on acylation of the hydroxy group is expected [26]. On the other hand, the discrimination of the chirality of rotaxane **10** is expected to be extremely difficult because the alcohol located at the edge of the molecule.

Catalyst **52** and other typical catalysts generally favor the sterically least congested conformer (Fig. 3.27b, gray-colored conformer). In this conformer, the ring component locates far from the reaction center of hydroxy group, thereby the hydroxy group is located in the least asymmetric microenvironments. Therefore, acylation of the hydroxy group of this conformer hardly results in effective kinetic resolution.

On the other hand, catalyst **1c** discriminate the mobile supramolecular chiral environments probably at the moment when the NHNs group and the hydroxy group

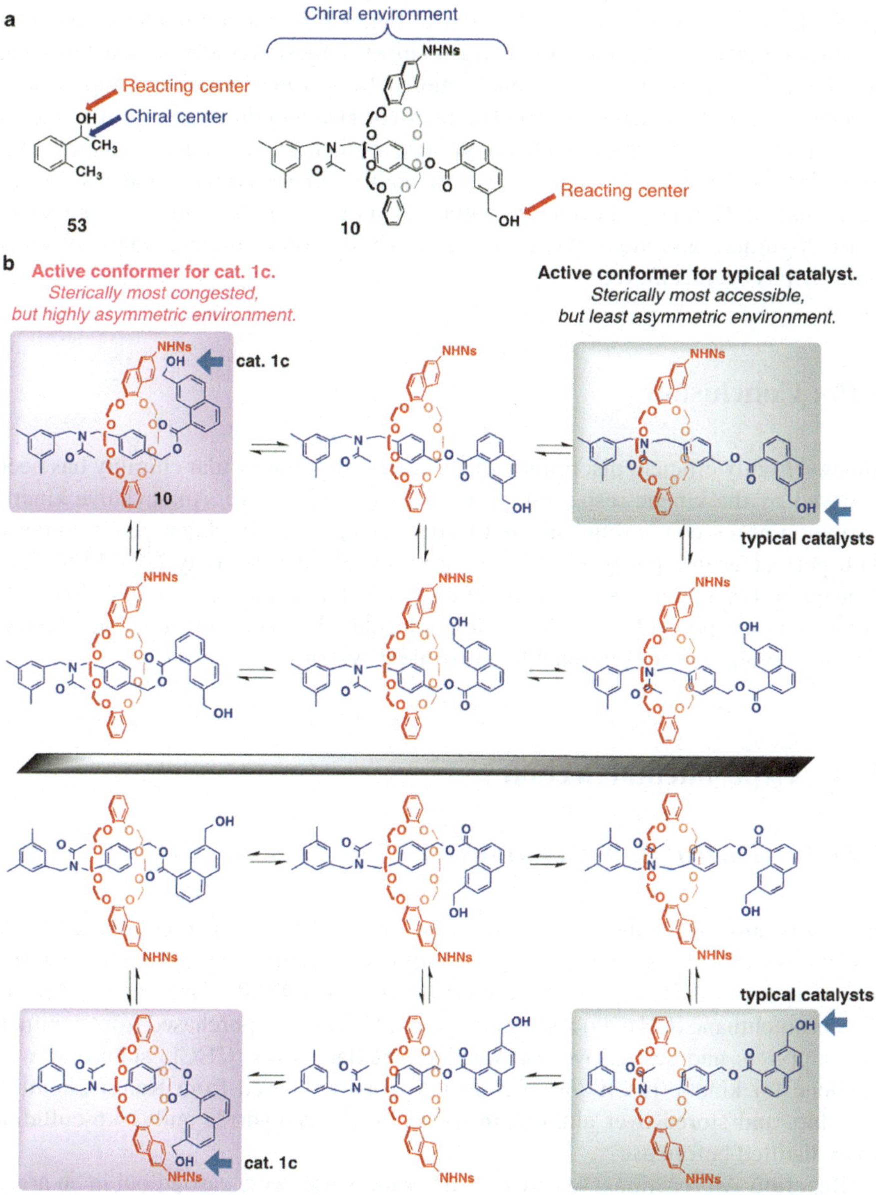

Fig. 3.27 Modes for the chiral discrimination. **a** Difference in the mode for the chiral discrimination between central chirality and mechanically planar chirality, **b** difference in the mode for the chiral discrimination between typical catalysts and catalyst **1c**

approach each other by around 5 Å distance (Fig. 3.27b, red-colored conformer). In this conformer, rotaxane **10** has the sterically most congested conformation, however, the chiral environment becomes much more obvious because ring component is located close to the reacting center. The hydroxy group in this active conformer for catalyst **1c** is located in the highly asymmetric microenvironments, although this conformer is the sterically least accessible conformer for typical catalysts. In this way, catalyst **1c** utilize the mobile nature of supramolecular chirality for molecular recognition, and the highly efficient discrimination of mobile supramolecular chirality has been achieved.

3.10 Conclusion

The first highly efficient discrimination of mobile supramolecular chirality has been achieved by the kinetic resolution promoted with catalyst **1c**. An acylative kinetic resolution of a racemic rotaxane **10** afforded a mechanically planar chiral rotaxane with perfect enantiopurity (>99% ee) in an excellent selectivity ($s > 14.5$). The concept of dynamic molecular recognition enabled asymmetric synthetic chemistry to set foot into the fields of mobile supramolecular chemistry linked a non-covalent system, beyond the conventional fixed covalent system.

3.11 Experimental Section

3.11.1 General Considerations

Reagents and solvents Unless otherwise noted, all commercial reagents and solvents were used as received. Anhydrous toluene, *N*,*N*-dimethylformamide (DMF), tetrahydrofuran (THF), pyridine, dimethylsulfoxide (DMSO), methanol (MeOH), dichloromehthane (CH_2Cl_2) and chloroform ($CHCl_3$) were purchased from commercial suppliers and stored over activated molecular sieves. $CHCl_3$ stabilized with amylene for kinetic resolution of rotaxanes was purchased from Kanto Chemical Co., Inc. and stored over aluminium oxide. Acetic anhydride and 2,4,6-collidine were distilled before use.

Reaction All reactions, sensitive to air or moisture, were carried out in an argon atmosphere under anhydrous conditions. Thin-layer chromatography (TLC) was performed on Silica gel 60 F_{254} precoated plates (0.25 mm, Merck). Visualization of developed chromatogram was accomplished with UV light and *p*-anysaldehyde (conc. H_2SO_4 in ethanol) or phosphomolybdic acid (5 w/v% in ethanol) stain followed by heating.

Purification Chromatographic purification of products was accomplished by using Silica gel 60 N (63–210 μm, Kanto Chemical Co., Inc.). Preparative TLC (PTLC) was carried out by using Silica gel 60 N F_{254} (0.5 mm, Merck).

Analysis Analytical HPLC was run on Waters 1525 Binary HPLC Pump, equipped with Waters 2998 Photodiode Array Detector. ^{1}H and ^{13}C NMR spectra were recorded on JEOL ECX-400 (400 and 100 MHz) and JEOL ECA-600 (600 and 150 MHz). Chemical shifts are reported relative to the solvent ($CHCl_3$: δ (^{1}H) = 7.26 ppm, δ (^{13}C) = 77.0 ppm) as reference. Data for ^{1}H NMR are reported as follows: chemical shift (δ ppm), integration, multiplicity (s = singlet, d = doublet, t = triplet, dd = double doublet, ddd = double double doublet, dt = double triplet, m = multiplet, br = broad, brt = broad triplet) and coupling constant (Hz). Data for proton-decoupled ^{13}C NMR are reported in terms of chemical shift. Infrared (IR) spectra were recorded on a JASCO FT-IR 4200 spectrometer and are reported in terms of frequency of absorption (cm^{-1}). High resolution mass spectra (HRMS) were obtained using JEOL JMS-700 mass spectrometer and Bruker Impact HD mass spectrometer. Melting points (m.p.) were recorded using Yanagimoto Micro Melting Point Apparatus PM-500. Specific rotations were measured with JASCO P-2200 polarimeter and HORIBA SEPA-200 automatic digital polarimeter, and are reported as follows: $[\alpha]_D{}^t$ (c = 10 mg/ml, solvent, enantiomeric excess (ee)). UV/Vis absorption spectra were recorded with a JASCO V-550 UV/Vis spectrophotometer. Circular dichroism (CD) spectra were recorded with a JASCO J-720W spectropolarimeter and a JASCO J-820-L spectropolarimeter equipped with a JASCO PTC-423L Peltier Controller.

3.11.2 Supplementary Methods

Procedures for preparation of enantiopure mechanically planar chiral rotaxane 10 (Fig. 3.24) To a $CHCl_3$ solution (2.25 ml) of catalyst **1c** (54.3 mg, 0.0635 mmol, 1.5 equivalents) were added a $CHCl_3$ solution of 2,4,6-collidine (0.075 M, 0.67 ml, 0.0503 mmol, 1.2 equivalents) and a $CHCl_3$ solution (1.0 ml) of *racemic* rotaxane **10** (50.0 mg, 0.0424 mmol, 1.0 equivalents) at room temperature. The mixture was cooled to −60 °C. A $CHCl_3$ solution of acetic anhydride (0.105 M, 0.32 ml, 0.0336 mmol, 0.8 equivalents) was added dropwise to the mixture, and the reaction mixture was stirred at −60 °C for 37 h. Then MeOH (15 ml) was added to the reaction mixture at −60 °C, and the mixture was stirred at room temperature for 30 min. The mixture was evaporated to dryness under a reduced pressure. The crude product was subjected to preparative TLC (SiO_2, MeOH/$CHCl_3$ 1/19) to allow isolation of recovered **10** (15.0 mg, 29% yield) and acylated **10-OAc** (33.0 mg, 65% yield). Enantiomeric purity of the recovered rotaxane, **10** (>99.9% ee), and the acylated product, **10-OAc** (40% ee), was determined by chiral HPLC with CHIRALPAK-IC column eluted by EtOH/CH_2Cl_2 1/60 (0.7 ml/min) at 20 °C. Conversion (C, C = 71%) and selectivity factor (s, s > 14.5) were determined according to the following equation: conversion $C = ee_{10}/(ee_{10} + ee_{10\text{-}OAc})$ and selectivity factor $s = \ln[(1-C)(1-ee_{10})]/\ln[(1-C)(1+ee_{10})] = \ln[1-C(1+ee_{10\text{-}OAc})]/\ln[1-C(1-ee_{10\text{-}OAc})]$.

General procedures for kinetic resolution of *racemic* rotaxane 10 on an analytical scale (Fig. 3.21, entry 5) To a $CHCl_3$ solution (15 μl) of catalyst **1c** (1.1 mg, 1.3 μmol, 1.5 equivalents), were added $CHCl_3$ (31 μl), a $CHCl_3$ solution of 2,4,6-collidine (0.075 M, 13.4 μl, 1.0 μmol, 1.2 equivalents) and a $CHCl_3$

solution (20 μl) of rotaxane **10** (1.0 mg, 0.85 μmol, 1.0 equivalents) at room temperature. The mixture was cooled to –60 °C. A $CHCl_3$ solution of acetic anhydride (0.105 M, 4.0 μl, 0.42 μmol, 0.5 equivalents) was added to the mixture and the reaction mixture was kept at −60 °C for 6 h. Then MeOH (300 μl) was added to the reaction mixture at −60 °C, and the mixture was kept at room temperature for 30 min. The mixture was evaporated to dryness under a reduced pressure. The resulting products were analyzed by chiral stationary phase HPLC with CHIRALPAK-IC column eluted by $EtOH/CH_2Cl_2 = 1/60$ (0.7 ml/min) at 20 °C to determine the enantiomeric purity of the recovered rotaxane, **10** (63% ee), and the acylated product, **10-OAc** (79% ee). Conversion (C, C = 44%) and selectivity factor (s, $s = 16.1$) were determined according to the following equation: conversion $C = ee_{10}/(ee_{10} + ee_{10\text{-OAc}})$ and selectivity factor $s = \ln[(1–C)(1–ee_{10})]/\ln[(1–C)(1 + ee_{10})] = \ln[1–C(1 + ee_{10\text{-OAc}})]/\ln[1–C(1–ee_{10\text{-OAc}})]$.

3.11.3 Analytical Data

3.11.3.1 Analytical Data of Rotaxane 10

See Fig. 3.28.

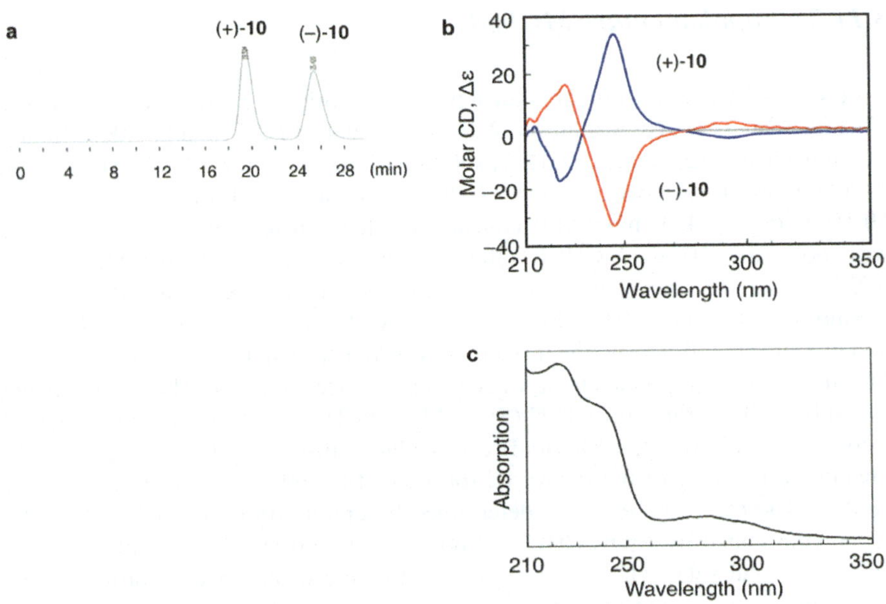

Fig. 3.28 Analytical data of rotaxane **10**. **a** Chiral HPLC chromatogram of a pair of enantiomers of rotaxane **10** (Conditions for HPLC: CHIRALPAK IC column (4.6 × 250 mm); eluent $EtOH/CH_2Cl_2$ 1/60; flow rate 0.7 ml/min; detection 254 nm, temperature 20 °C), **b** CD spectra of a pair of enantiomers of rotaxane **10** (Recorded in 0.5×10^{-5} M MeOH), **c** UV/Vis spectrum of **10** (Recorded in 0.5×10^{-5} M MeOH)

3.11.3.2 Chiral HPLC Analysis of Kinetic Resolution of Rotaxanes 10, 14 and 15

See Figs. 3.29, 3.30, 3.31 and 3.32.

a

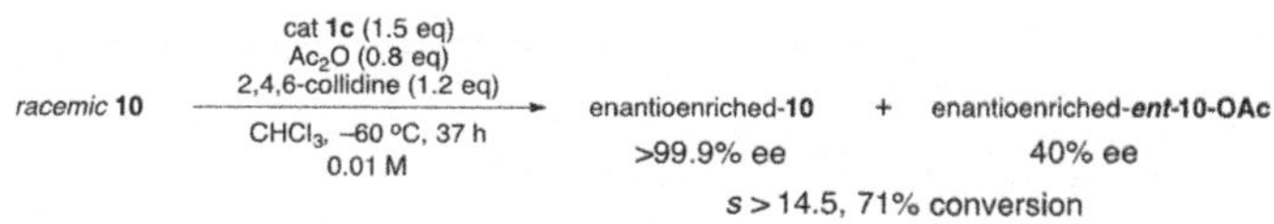

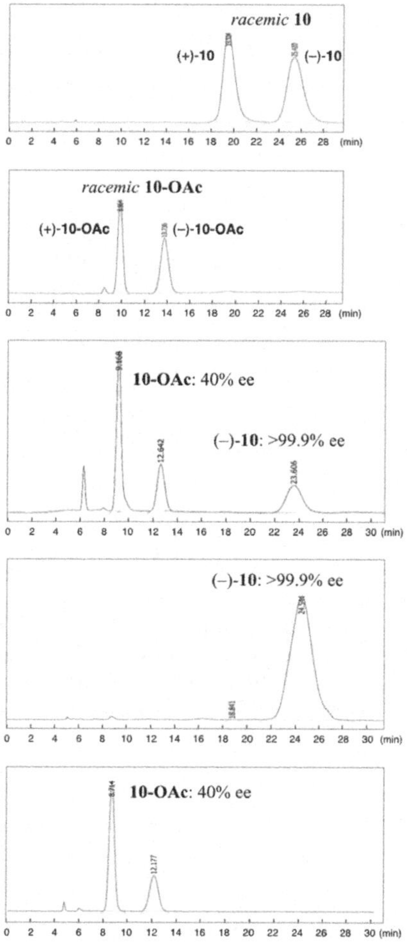

b

Retention time (min)	Area	% Area
19.524	3765444	50.34
25.420	3714530	49.66

c

Retention time (min)	Area	% Area
9.864	6906524	50.37
13.736	6806010	49.63

d

Retention time (min)	Area	% Area
9.168	11240033	51.09
12.642	4831686	21.96
23.606	5928000	26.95

e

Retention time (min)	Area	% Area
18.841	10026	0.01
24.544	67024334	99.99

f

Retention time (min)	Area	% Area
8.714	13329617	69.98
12.177	5719404	30.02

Fig. 3.29 Analysis of kinetic resolution of *racemic* **10** shown in Fig. 3.24. **a** Reaction scheme, Chiral HPLC chromatograms of **b** *racemic* **10**, **c** *racemic* **10-OAc**, **d** crude product of the kinetic resolution, **e** isolated enantioenriched **10**, and **f** isolated enantioenriched **10-OAc** (Conditions for HPLC: CHIRALPAK IC column (4.6 × 250 mm); eluent 1/60 EtOH/CH_2Cl_2; flow rate 0.7 ml/min; detection 254 nm, temperature 20 °C)

a

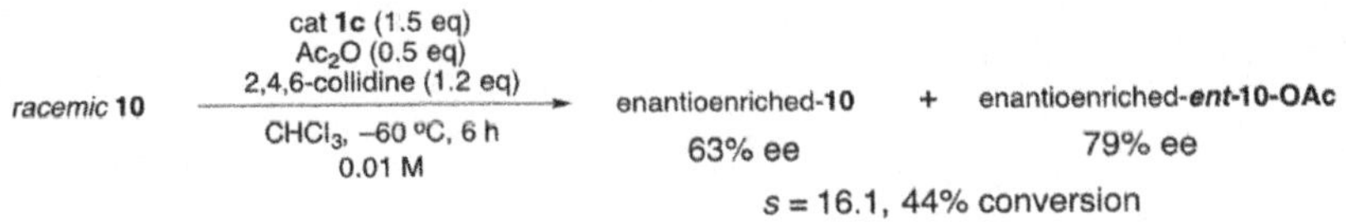

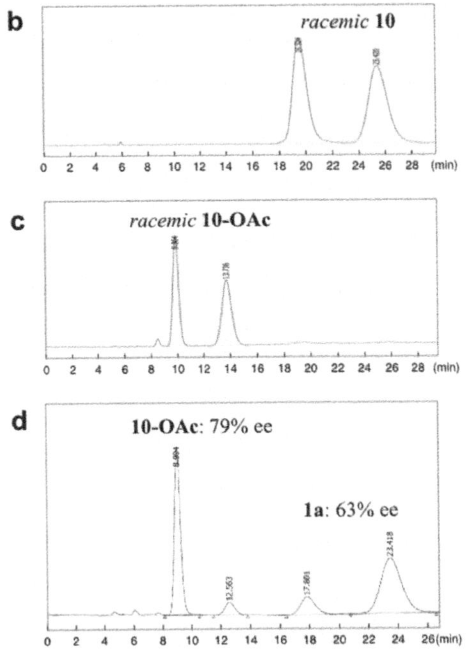

Retention time (min)	Area	% Area
19.524	3765444	50.34
25.420	3714530	49.66

Retention time (min)	Area	% Area
9.864	6906524	50.37
13.736	6806010	49.63

Retention time (min)	Area	% Area
8.994	4826739	40.71
12.563	569000	4.80
17.801	1191675	10.05
23.418	5267915	44.43

Fig. 3.30 Analysis of kinetic resolution of *racemic* **10** shown in Fig. 3.21, entry 5. **a** Reaction scheme, Chiral HPLC chromatograms of **b** *racemic* **10**, **c** *racemic* **10-OAc**, and **d** crude product of the kinetic resolution of **10** (Conditions for HPLC: CHIRALPAK IC column (4.6 × 250 mm); eluent EtOH/CH_2Cl_2 1/60; flow rate 0.7 ml/min; detection 254 nm, temperature 20 °C)

a

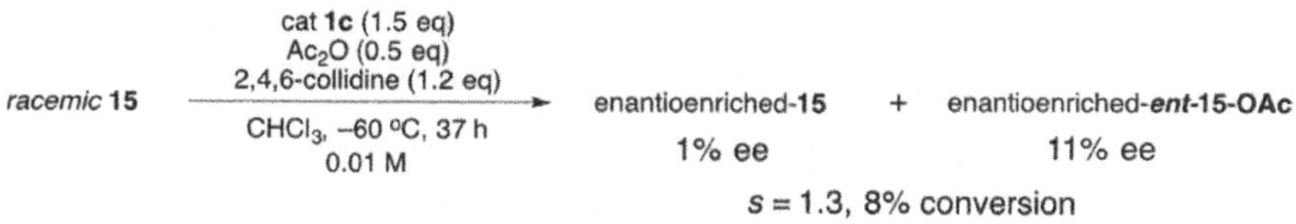

b

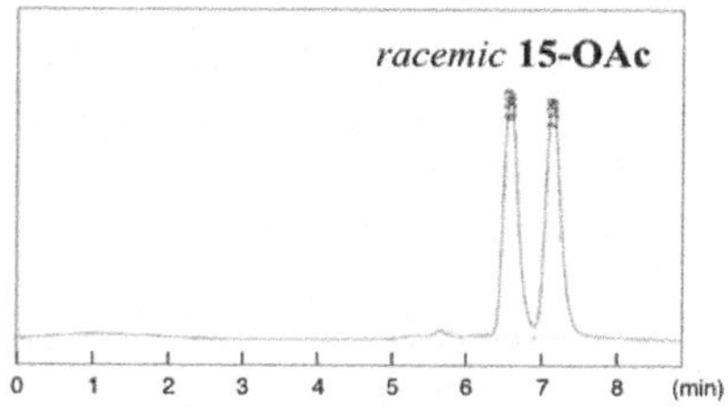

Retention time (min)	Area	% Area
6.567	15496078	49.37
7.129	15888699	50.63

c

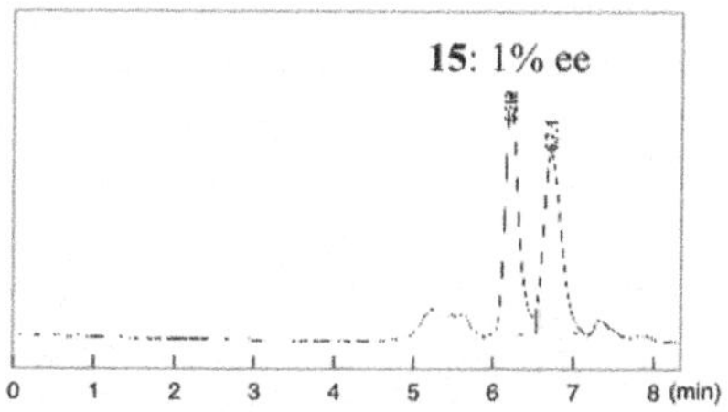

Retention time (min)	Area	% Area
6.212	94705	50.68
6.714	92180	49.32

d

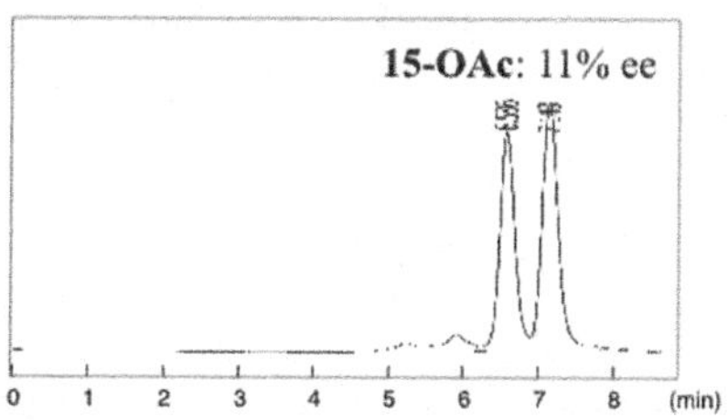

Retention time (min)	Area	% Area
6.586	2505506	44.72
7.413	3097666	55.28

Fig. 3.31 Analysis of kinetic resolution of *racemic* **15**. **a** Reaction scheme, Chiral HPLC chromatograms of **b** *racemic* **15-OAc**, **c** enantioenriched **15** (ee of **15** was determined after converting to **15-OAc**), and **d** enantioenriched **15-OAc** (Conditions for HPLC: CHIRALPAK ID column (4.6 × 250 mm); eluent EtOH/$CHCl_3$ 1/540; flow rate 0.7 ml/min; temperature 20 °C)

a

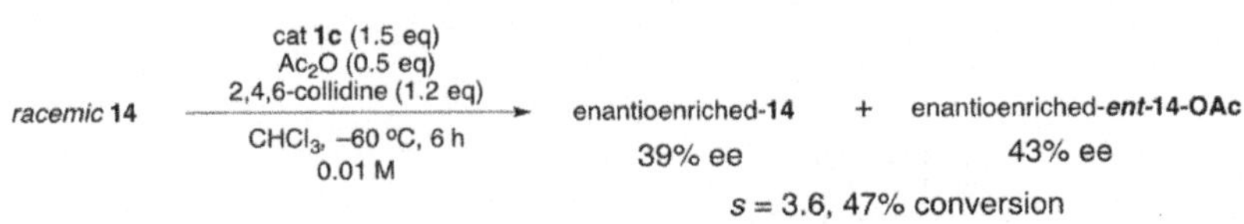

b

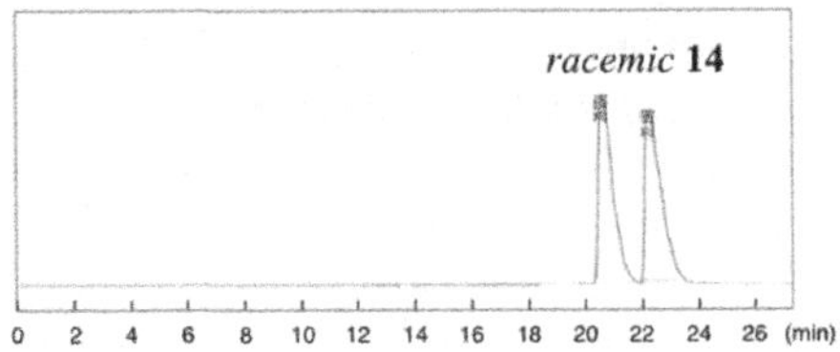

Retention time (min)	Area	% Area
20.551	108720926	49.17
22.193	112404403	50.83

c

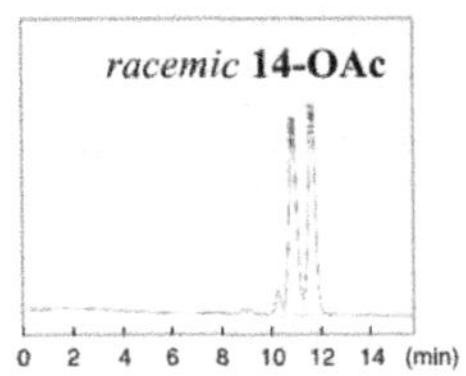

Retention time (min)	Area	% Area
10.850	5923032	50.26
11.617	5862701	49.74

d

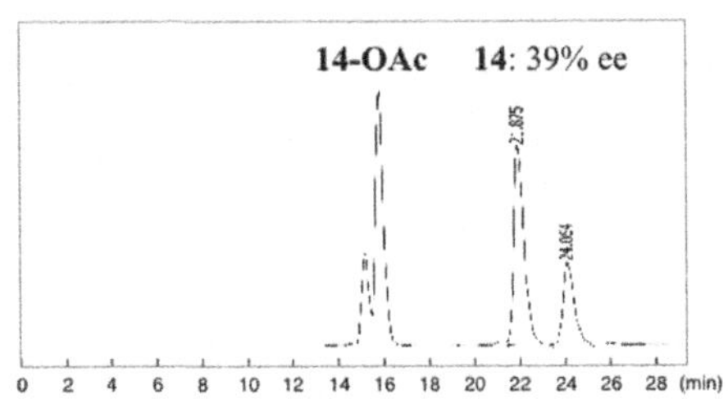

Retention time (min)	Area	% Area
21.875	30492692	69.54
24.064	13354707	30.46

e

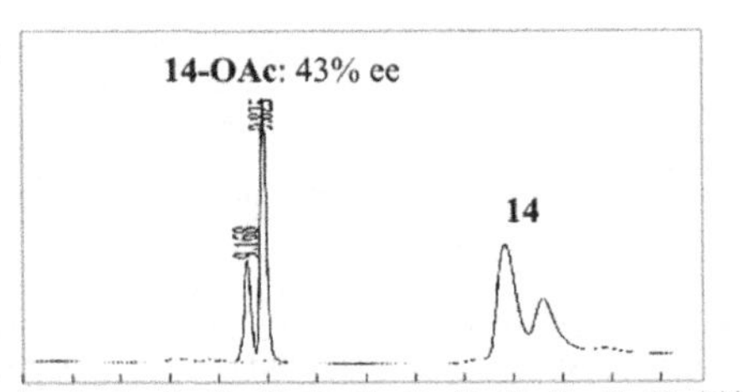

Retention time (min)	Area	% Area
9.168	686190	28.46
9.825	1725002	71.54

Fig. 3.32 Analysis of kinetic resolution of *racemic* **14**. **a** Reaction scheme, Chiral HPLC chromatograms of **b** *racemic* **14**, **c** *racemic* **14-OAc**, and **d**, **e** crude product of the kinetic resolution (Conditions for HPLC (**b**, **d**): CHIRALPAK ID column (4.6 × 250 mm); eluent EtOH/CH_2Cl_2 1/40; flow rate 0.5 ml/min; temperature 20 °C. Conditions for HPLC (**c**, **e**): CHIRALPAK ID column (4.6 × 250 mm); eluent EtOH/CH_2Cl_2 1/80; flow rate 1.0 ml/min; temperature 20 °C)

3.11.4 Synthesis of Compounds

3.11.4.1 Purification of Catalyst 1c

Catalyst **1c** was prepared according to the reported method [27], then further purification was performed in the following way. After extraction and evaporation of the reaction mixture, the residue was subjected to column chromatography (SiO_2, AcOEt/Hexane 2/1 to MeOH/$CHCl_3$ 1/99 to 1/50), affording a pale yellow residue containing **1c**. The pale yellow residue was dissolved in MeOH, and the mixture was evaporated to dryness under a reduced pressure. Then the resulting residue was dissolved in AcOEt, and the mixture was evaporated to dryness under a reduced pressure. The pale yellow residue was recrystallized from hot AcOEt by cooling and adding a small amount of hexane, affording a slightly pale yellow residue. To the slightly pale yellow residue was added a small amount of MeOH, which was filtered to afford completely pure catalyst **1c** (7.02 g) as white solid.

3.11.4.2 Synthesis of Ring Components 20–22

6-Bromonaphthalene-1,2-dione (24) To a THF solution (960 ml) of 6-bromo-2-naphthol (35.6 g, 160 mmol) was added 2-iodoxybenzoic acid (50 g, 178 mmol) at room temperature, which changed its color to bright orange within 1 h. The reaction mixture was stirred at room temperature for 16 h in the dark. The reaction mixture was evaporated to dryness under a reduced pressure. The residue was diluted with $CHCl_3$. The mixture was washed successively with saturated aqueous solution of $NaHCO_3$ and brine, and an organic phase separated was dried over Na_2SO_4, filtered and evaporated to dryness under a reduced pressure. The residue was obtained as bright orange solid, which was directly used for the next step.

6-Bromonaphthalene-1,2-diol (25) The following steps were carried out under Ar atmosphere. All solvents and reaction vessels were degassed by flushing with Ar gas. The bright orange solid containing 6-bromonaphthalene-1,2-dione (**24**) was dissolved in degassed THF (245 ml) at room temperature under Ar. To the solution was added a degassed aqueous solution (222 ml) of $Na_2S_2O_4$ (145 g, 835 mmol), then the mixture changed its color to pale orange within a few minutes. The reaction mixture was stirred under Ar atmosphere at room temperature for 15 min. The mixture was evaporated to dryness under a reduced pressure under Ar atmosphere. The residue was directly used for the next step.

Compound 26 The following steps were carried out under Ar atmosphere. All solvents and reaction vessels were degassed by flushing with Ar gas. To a degassed MeCN solution (50 ml) of a mixture of triethylene glycol monotosylate (52 g, 171 mmol) and powdered K_2CO_3 (66.4 g, 480 mmol) was added a degassed MeCN solution (300 ml) of 6-bromonaphthalene-1,2-diol (**25**) at reflux temperature under Ar atmosphere. The reaction mixture was refluxed under Ar atmosphere for 37 h. The reaction mixture was cooled to room temperature, filtered through a celite pad, and the filtrate was evaporated to dryness under a reduced pressure. The residue was subjected to column chromatography (SiO_2, Hexane/Acetone 1/1) to allow isolation of **26** (20.9 g, 26% yield, 3 steps) as pale yellow oil. ^{1}H NMR (600 MHz, $CDCl_3$): δ 8.09 (d, J = 8.9 Hz, 1H), 7.90 (d, J = 1.4 Hz, 1H), 7.50 (dd, J = 8.9, 2.0 Hz, 1H), 7.45 (d, J = 8.9 Hz, 1H), 7.27 (d, J = 9.6 Hz, 1H), 4.35 (t, J = 4.8 Hz, 2H), 4.29 (t, J = 4.8 Hz, 2H), 3.88 (t, J = 4.8 Hz, 2H), 3.86 (t, J = 4.8 Hz, 2H), 3.74–3.67 (m, 12H), 3.61–3.58 (m, 4H), 2.95 (br, 2H). ^{13}C NMR (150 MHz, $CDCl_3$): δ 147.38, 142.57, 130.86, 129.24, 129.15, 127.88, 123.71, 123.05, 118.25, 118.18, 72.52, 72.49, 72.45, 70.64, 70.47, 70.44, 70.32, 70.28, 69.80, 69.43, 61.54. IR (neat): 3384, 2873, 1587, 1496, 1456, 1348, 1272, 1105, 1070 cm^{-1}. HRMS (EI): m/z calcd for $C_{22}H_{31}{}^{79}BrO_8$ $[M]^+$: 502.1202, found: 502.1206; calcd for $C_{22}H_{31}{}^{81}BrO_8$ $[M]^+$: 504.1185, found: 504.1186.

Compound 27 To a mixture of a CH_2Cl_2 solution (100 ml) of **26** (20.1 g, 40 mmol), triethylamine (56 ml, 400 mmol), and a CH_2Cl_2 solution (100 ml) of *N*,*N*-dimethyl-4-aminopyridine (970 mg, 8.0 mmol) was dropwisely added a CH_2Cl_2 solution (250 ml) of *p*-toluenesulfonyl chloride (33.6 g, 176 mmol) over a period of 2 h at room temperature. The reaction mixture was stirred at room temperature for 26 h, which changed its color to reddish brown. Then 5 M aqueous HCl solution was carefully added to the reaction mixture, and the mixture was extracted with CH_2Cl_2. The combined organic extract was washed successively with 2 M aqueous HCl solution, water and brine, and an organic phase separated was dried over Na_2SO_4, filtered, and evaporated to dryness under a reduced pressure. The residue was subjected to column chromatography (SiO_2, Hexane/AcOEt 1/1 to 0/1) to allow isolation of **27** (22.0 g, 68% yield) as orange oil. ^{1}H NMR (600 MHz, $CDCl_3$): δ 8.08 (d, J = 8.9 Hz, 1H), 7.89 (d, J = 1.4 Hz, 1H), 7.79–7.75 (m, 4H), 7.47–7.44 (m, 2H), 7.32–7.29 (m, 4H), 7.27 (d, J = 8.9 Hz, 1H), 4.31 (t, J = 4.8 Hz, 2H), 4.27 (t, J = 4.8 Hz, 2H), 4.15–4.12 (m, 4H), 3.84 (t, J = 4.8 Hz, 2H), 3.77 (t, J = 4.8 Hz, 2H), 3.69–3.59 (m, 12H), 2.41 (s, 3H), 2.40 (s, 3H). ^{13}C NMR (150 MHz, $CDCl_3$): δ 147.42, 144.70, 144.67, 142.46, 132.72, 130.76, 129.69, 129.14, 128.99, 127.91, 127.76, 123.80, 122.94, 118.28, 118.07, 72.31, 70.61, 70.50, 70.36, 70.30, 69.84, 69.39, 69.16, 69.13, 68.56, 21.45. IR (neat): 2875, 1589, 1496, 1453, 1354, 1176, 1127, 1098 cm^{-1}. HRMS (FAB): m/z calcd for $C_{36}H_{43}{}^{79}BrO_{12}S_2$ $[M]^+$: 810.1379, found: 810.1408; calcd for $C_{36}H_{43}{}^{81}BrO_{12}S_2$ $[M]^+$: 812.1365, found: 812.1371.

Compound 28 A MeCN solution (1000 ml) of a mixture of catechol (3.0 g, 27 mmol) and Cs_2CO_3 (44.0 g, 135 mmol) was refluxed for 15 min. To the mixture was dropwisely added a MeCN solution (268 ml) of **27** (21.9 g, 27 mmol) over a period of 1 h at reflux temperature. The reaction mixture was refluxed for 7 days. The mixture was evaporated to dryness under a reduced pressure, and the residue was diluted with $CHCl_3$. The mixture was washed successively with saturated aqueous solution of $NaHCO_3$ and brine, and an organic phase separated was dried over Na_2SO_4, filtered and evaporated to dryness under a reduced pressure. The residue was subjected to column chromatography (SiO_2, MeOH/$CHCl_3$ 1/99), affording oil product. The oil product was subjected to column chromatography (SiO_2, $CHCl_3$/Hexane/AcOEt

1/1/4) to allow isolation of **28** (10.0 g, 64% yield) as pale yellow solid. M.p. 58–61 °C. ^{1}H NMR (600 MHz, $CDCl_3$): δ 8.02 (d, J = 8.9 Hz, 1H), 7.90 (d, J = 1.4 Hz, 1H), 7.49 (dd, J = 8.9, 2.1 Hz, 1H), 7.45 (d, J = 8.9 Hz, 1H), 7.25 (d, J = 8.9 Hz, 1H), 6.90–6.87 (m, 4H), 4.34 (t, J = 4.8 Hz, 2H), 4.29 (t, J = 4.4 Hz, 2H), 4.16–4.14 (m, 4H), 3.95 (t, J = 4.8 Hz, 2H), 3.93–3.90 (m, 6H), 3.86–3.83 (m, 4H), 3.82–3.79 (m, 4H). ^{13}C NMR (150 MHz, $CDCl_3$): δ 148.91, 148.84, 147.60, 142.51, 130.70, 129.35, 129.18, 127.77, 123.58, 123.01, 121.47, 121.39, 118.01, 117.46, 114.35, 114.08, 72.39, 71.09, 71.06, 70.62, 70.58, 69.99, 69.89, 69.83, 69.51, 69.32, 69.10. IR (KBr): 3427, 2922, 2886, 1622, 1589, 1501, 1450, 1360, 1329, 1269, 1206, 1135, 1070, 745 cm^{-1}. HRMS (FAB): m/z calcd for $C_{28}H_{33}{}^{79}BrO_8$ [M]$^+$: 576.1359, found: 576.1351; calcd for $C_{28}H_{33}{}^{81}BrO_8$ [M]$^+$: 578.1343, found: 578.1341.

Compound 29 To a degassed toluene solution (55 ml) of **28** (7.97 g, 13.8 mmol) was added successively $Pd_2(dba)_3$ (885 mg, 0.97 mmol), ^{t}BuXPhos (1.39 g, 2.9 mmol), NaO^{t}Bu (1.86 g, 19.3 mmol) and *tert*-butyl carbamate (1.94 g, 16.6 mmol) at room temperature. After being stirred at room temperature for 8 h, the reaction mixture was filtered through a celite pad, and evaporated to dryness under a reduced pressure. The residue was subjected to column chromatography (SiO_2, MeOH/$CHCl_3$ 1/99) to allow isolation of **29** (9.5 g, 99% yield) as red oil. ^{1}H NMR (600 MHz, $CDCl_3$): δ 8.05 (d, J = 8.9 Hz, 1H), 7.96 (br, 1H), 7.46 (d, J = 8.9 Hz, 1H), 7.24 (dd, J = 8.9, 2.0 Hz, 1H), 7.20 (d, J = 8.9 Hz, 1H), 6.90–6.87 (m, 4H), 6.64 (br, 1H), 4.32 (t, J = 4.8 Hz, 2H), 4.27 (t, J = 4.1 Hz, 2H), 4.16–4.14 (m, 4H), 3.95 (t, J = 4.8 Hz, 2H), 3.93–3.91 (m, 6H), 3.85–3.84 (m, 4H), 3.82–3.79 (m, 4H), 1.54 (s, 9H). ^{13}C NMR (150 MHz, $CDCl_3$): δ 152.84, 148.88, 148.81, 146.26, 142.48, 134.45, 130.33, 125.66, 123.29, 122.45, 121.41, 121.34, 119.44, 117.34, 114.34, 114.10, 80.34, 72.26, 70.98, 70.57, 70.52, 70.03, 69.80, 69.74, 69.63, 69.27, 69.09, 28.25. IR (neat): 3319, 2929, 2873, 1719, 1604, 1543, 1502, 1454, 1369, 1250, 1158, 1128, 1058, 750 cm^{-1}. HRMS (FAB): m/z calcd for $C_{33}H_{43}NNaO_{10}$ [M + Na]$^+$: 636.2785, found: 636.2781; calcd for $C_{33}H_{43}NO_{10}$ [M]$^+$: 613.2887, found: 613.2893.

Compound 30 4 M dioxane solution of HCl (80 ml) was added to a reaction vessel containing **29** (9.9 g, 16.1 mmol) at room temperature. After being stirred at room temperature for 7 h, the mixture was evaporated to dryness under a reduced pressure. The residue was directly used for the next step.

Ring component 20 To a MeOH/water (161 ml/161 ml) solution of a mixture of **30** and NaOAc (7.4 g, 90 mmol) was added 2-nitrobenzenesulfonyl chloride (12.5 g, 56 mmol) at room temperature. The reaction mixture was stirred at 50 °C for 12 h. Then $CHCl_3$ was added to the reaction mixture, and the mixture was extracted with $CHCl_3$. The combined organic extract was washed successively with saturated aqueous solution of $NaHCO_3$ and brine, and an organic phase separated was dried over Na_2SO_4, filtered and evaporated to dryness under a reduced pressure. The residue was subjected to column chromatography (SiO_2, MeOH/AcOEt 1/99) to allow isolation of **20** (9.69 g, 86% yield, 2 steps) as pale yellow amorphous. ^{1}H NMR (600 MHz, $CDCl_3$): δ 8.04 (d, J = 8.9 Hz, 1H), 7.85 (d, J = 7.6 Hz, 1H), 7.78 (dd, J = 7.5, 1.3 Hz, 1H), 7.66–7.62 (m, 1H), 7.57 (d, J = 2.0 Hz, 1H), 7.51–7.47 (m, 1H), 7.44 (d, J =

8.9 Hz, 1H), 7.36 (br, 1H), 7.25–7.22 (m, 2H), 6.90–6.86 (m, 4H), 4.31 (t, $J = 4.8$ Hz, 2H), 4.27 (t, $J = 4.8$ Hz, 2H), 4.15–4.12 (m, 4H), 3.93 (t, $J = 5.5$ Hz, 2H), 3.91–3.89 (m, 6H), 3.84–3.82 (m, 4H), 3.80–3.77 (m, 4H). ^{13}C NMR (150 MHz, $CDCl_3$): δ 148.84, 148.72, 148.02, 147.46, 142.24, 133.85, 132.48, 131.99, 131.66, 131.36, 129.55, 127.55, 125.09, 123.58, 123.30, 122.26, 121.44, 121.31, 120.77, 117.40, 114.34, 113.96, 72.30, 70.94, 70.51, 70.46, 69.87, 69.78, 69.69, 69.44, 69.24, 68.98. IR (neat): 2927, 2874, 1600, 1543, 1503, 1453, 1370, 1255, 1124, 748 cm^{-1}. HRMS (FAB): *m*/*z* calcd for $C_{34}H_{38}N_2O_{12}SNa$ $[M + Na]^+$: 721.2043, found: 721.2037.

3.11.4.3 Synthesis of Endcap Component 23

1,7-Naphthylbis(trifluoromethanesulfonate) (33) To a pyridine solution (200 ml) of 1,7-dihydroxynaphthalene (23.4 g, 146 mmol) was dropwisely added trifluoromethanesulfonic anhydride (49 ml, 292 mmol) at 0 °C. The reaction mixture was allowed to warm to room temperature, and stirred for 3 h. Water was added to the reaction mixture, and the mixture was extracted with AcOEt. The combined organic extract was washed successively with 1 M aqueous HCl solution, water and brine, and an organic phase separated was dried over Na_2SO_4, filtered and evaporated to dryness under a reduced pressure. The residue was subjected to column chromatography (SiO_2, Hexane/AcOEt 5/1) to allow isolation of 1,7-naphthylbis(trifluoromethanesulfonate) (**33**) (61.3 g, 99% yield) as colorless oil. ^{1}H NMR (600 MHz, $CDCl_3$): δ 8.04 (d, $J = 8.9$ Hz, 1H), 7.94–7.97 (m, 2H), 7.63–7.60 (m, 2H), 7.53 (dd, $J = 8.9$, 2.7 Hz, 1H). ^{13}C NMR (150 MHz, $CDCl_3$): δ 148.43, 145.12, 133.59, 131.07, 128.33, 126.65, 126.42, 121.38, 119.61, 118.85 (q, $J =$ 318.8 Hz), 118.76 (q, $J = 318.8$ Hz), 112.89. IR (neat): 1606, 1510, 1428, 1215, 1138, 1204 cm^{-1}. HRMS (EI): *m*/*z* calcd for $C_{12}H_6F_6O_6S_2$ $[M]^+$: 423.9510, found: 423.9494.

Dimethyl naphthalene-1,7-dicarboxylate (34) To a DMSO/MeOH (153 ml/153 ml) solution of 1,7-naphthylbis(trifluoromethanesulfonate) (**33**) (18.6 g, 43.8 mmol) was added triethylamine (92 ml, 660 mmol) followed by $Pd(OAc)_2$ (1.1 g, 4.38 mmol) and 1,3-Bis(diphenylphosphino)propane (1.8 g, 4.38 mmol) at room temperature. A stream of CO was passed into the solution for 2–3 min, then the reaction vessel and contents were placed in a 55–65 °C oil bath with a CO balloon. After being stirred for 36 h, the reaction mixture was filtered through a celite pad, and a part of solvent was evaporated under a reduced pressure. The mixture was diluted with AcOEt, and extracted with AcOEt. The combined organic extract was washed successively with 1 M aqueous HCl solution, water and brine, and an organic phase separated was dried over Na_2SO_4, filtered and evaporated to dryness under a reduced pressure. The residue was subjected to column chromatography (SiO_2, Hexane/AcOEt 5/1) to allow isolation of dimethyl naphthalene-1,7-dicarboxylate (**34**) (3.47 g, 32% yield) as yellow solid. M.p. 83–85 °C. ^{1}H NMR (600 MHz, $CDCl_3$): δ 9.65 (s, 1H), 8.23 (d, $J = 6.2$ Hz, 1H), 8.13

(dd, $J = 8.9$, 1.4 Hz, 1H), 8.04 (d, $J = 8.3$ Hz, 1H), 7.92 (d, $J = 8.3$ Hz, 1H), 7.60 (t, $J = 8.3$ Hz, 1H), 4.03 (s, 3H), 3.99 (s, 3H). ^{13}C NMR (150 MHz, $CDCl_3$): δ 167.27, 167.08, 135.56, 132.87, 130.69, 130.23, 128.84, 128.60, 128.00, 126.58, 125.52, 52.21. IR (KBr): 2949, 1717, 1457, 1435, 1272, 1242, 1196, 1146 cm^{-1}. HRMS (EI): *m/z* calcd for $C_{14}H_{12}O_4$ $[M]^+$: 244.0736, found: 244.0740.

8-(Methoxycarbonyl)-2-naphthoic acid (35) To a dioxane solution (360 ml) of dimethyl naphthalene-1,7-dicarboxylate (**34**) (20.0 g, 81.9 mmol) was added a MeOH solution (17 ml) of KOH (4.73 g, 84.3 mmol) at room temperature. The reaction mixture was stirred at room temperature for 2 h. Water was added to the reaction mixture, and the mixture was extracted with AcOEt. The combined organic extract containing dimethyl naphthalene-1,7-dicarboxylate (**34**) was washed with water, and evaporated under a reduced pressure. This recovered dimethyl naphthalene-1,7-dicarboxylate (**34**) was again used for the above reaction. After repeating above reaction three times, the combined aqueous extract was acidified with 1 M aqueous HCl solution to pH 3. The aqueous solution was extracted with AcOEt. The combined organic extract was washed with brine, and an organic phase separated was dried over dried over Na_2SO_4, filtered and evaporated to dryness under a reduced pressure. The residue was subjected to column chromatography (SiO_2, MeOH/$CHCl_3$ 1/99), affording a mixture of 8-(methoxycarbonyl)-2-naphthoic acid (**35**) and 7-(methoxycarbonyl)-1-naphthoic acid. The mixture was recrystallized from $CHCl_3$ to give 8-(methoxycarbonyl)-2-naphthoic acid (**35**) (5.9 g, 31% yield) as white solid. M.p. 229–231 °C. ^1H NMR (600 MHz, $CDCl_3$): δ 9.77 (s, 1H), 8.27 (dd, $J = 6.8$, 1.4 Hz, 1H), 8.18 (dd, $J = 8.9$, 1.4 Hz, 1H), 8.09 (d, $J = 8.3$ Hz, 1H), 7.98 (d, $J = 8.9$ Hz, 1H), 7.65 (t, $J = 6.8$ Hz, 1H), 4.06 (s, 3H). ^{13}C NMR (150 MHz, $CDCl_3$): δ 170.15, 167.46, 136.16, 133.03, 130.95, 130.38, 129.99, 128.95, 128.59, 127.97, 127.19, 125.87, 52.47. IR (neat): 3411, 2950, 1713, 1510, 1454, 1436, 1281, 1251, 1226, 1136 cm^{-1}. HRMS (EI): *m/z* calcd for $C_{13}H_{10}O_4$ $[M]^+$: 230.0579, found: 230.0579.

Methyl 7-(hydroxymethyl)-1-naphthoate (36) To a THF solution (50 ml) of 8-(methoxycarbonyl)-2-naphthoic acid (**35**) (5.8 g, 25.3 mmol) was added a 1.0 M THF solution of borane tetrahydrofuran complex at 0 °C. The reaction mixture was stirred at 0 °C for 6 h. 1 M aqueous HCl solution was dropwisely added to the reaction mixture, and the mixture was extracted with AcOEt. The combined organic extract was washed successively with water and brine, and an organic phase separated was dried over Na_2SO_4, filtered and evaporated to dryness under a reduced pressure. The residue was subjected to column chromatography (SiO_2, MeOH/$CHCl_3$ 1/50 to 1/20) to allow isolation of methyl 7-(hydroxymethyl)-1-naphthoate (**36**) (4.6 g, 83% yield) as colorless oil. ^1H NMR (600 MHz, $CDCl_3$): δ 8.86 (s, 1H), 8.17 (d, $J = 7.6$ Hz, 1H), 8.00 (d, $J = 8.3$ Hz, 1H), 7.87 (d, $J = 8.2$ Hz, 1H), 7.56 (d, $J = 8.2$ Hz, 1H), 7.48 (t, $J = 8.2$ Hz, 1H), 4.88 (s, 2H), 3.99 (s, 3H), 2.12 (br, 1H). ^{13}C NMR (150 MHz, $CDCl_3$): δ 167.98, 140.41, 133.07, 131.10, 130.29, 128.74, 126.66, 125.42, 124.24, 123.05, 65.39, 52.08. IR (KBr): 3411, 3056, 2950, 1711, 1685, 1622, 1461, 1300, 1274, 1249, 1196, 1157, 1111 cm^{-1}. HRMS (EI): *m/z* calcd for $C_{13}H_{12}O_3$ $[M]^+$: 216.0786, found: 216.0785.

Compound 37 To a CH_2Cl_2 solution (200 ml) of methyl 7-(hydroxymethyl)-1-naphthoate (**36**) (4.54 g, 21 mmol) and 3,4-dihydro-2*H*-pyran (14.0 ml, 154 mmol) was added pyridinium *p*-toluenesulfonate (503 mg, 2.0 mmol) at room temperature. The reaction mixture was stirred at room temperature for 3 h. Saturated aqueous solution of $NaHCO_3$ was added to the reaction mixture, and the mixture was extracted with CH_2Cl_2. The combined organic extract was washed successively with water and brine, and an organic phase separated was dried over Na_2SO_4, filtered and evaporated to dryness under a reduced pressure. The residue was subjected to column chromatography (SiO_2, Hexane/AcOEt 5/1) to allow isolation of **37** (6.3 g, 99% yield) as colorless oil. ^{1}H NMR (600 MHz, $CDCl_3$): δ 8.88 (s, 1H), 8.18 (dd, J = 7.6, 1.4 Hz, 1H), 8.01 (d, J = 8.2 Hz, 1H), 7.88 (d, J = 8.3 Hz, 1H), 7.59 (dd, J = 8.3, 1.4 Hz, 1H), 7.48 (t, J = 7.6 Hz, 1H), 4.99 (d, J = 12.4 Hz, 1H), 4.78 (t, J = 3.4 Hz, 1H), 4.72 (d, J = 12.4 Hz, 1H), 4.00 (s, 3H), 4.00–3.95 (m, 1H), 3.61–3.56 (m, 1H), 1.93–1.87 (m, 1H), 1.80–1.75 (m, 1H), 1.73–1.68 (m, 1H), 1.64–1.53 (m, 3H). ^{13}C NMR (150 MHz, $CDCl_3$): δ 167.81, 137.78, 133.18, 132.99, 131.11, 130.25, 128.61, 126.80, 126.09, 124.25, 97.73, 69.05, 62.01, 51.96, 30.46, 25.35, 19.68. IR (neat): 2947, 1717, 1510, 1455, 1437, 1281, 1251, 1135, 1034 cm^{-1}. HRMS (EI): *m*/*z* calcd for $C_{18}H_{20}O_4$ $[M]^+$: 300.1362, found: 300.1366.

Endcap component 23 To a THF/MeOH (210 ml/84 ml) solution of **37** (6.3 g, 20.8 mmol) was added an aqueous solution (42 ml) of KOH (5.3 g, 94 mmol) at room temperature. The reaction mixture was stirred at room temperature for 12 h. A part of solvent was evaporated under a reduced pressure, and the mixture was diluted with $CHCl_3$. The mixture was partitioned between water and $CHCl_3$. The aqueous phase separated was neutralized with an aqueous solution of citric acid, and the mixture was extracted with $CHCl_3$. The combined organic extract was washed successively with water and brine, and an organic phase separated was dried over Na_2SO_4, filtered and evaporated to dryness under a reduced pressure. The residue was subjected to column chromatography (SiO_2, MeOH/AcOEt 1/99) to allow isolation of **23** (5.0 g, 84% yield) as white solid. M.p. 86–89 °C. ^{1}H NMR (600 MHz, $CDCl_3$): δ 9.06 (s, 1H), 8.39 (dd, J = 8.2, 1.4 Hz, 1H), 8.07 (d, J = 8.4 Hz, 1H), 7.90 (d, J = 8.3 Hz, 1H), 7.61 (dd, J = 8.9, 1.4 Hz, 1H), 7.53 (t, J = 7.6 Hz, 1H), 5.02 (d, J = 12.4 Hz, 1H), 4.81 (t, J = 3.4 Hz, 1H), 4.76 (d, J = 12.4 Hz, 1H), 3.99 (ddd, J = 11.0, 8.9, 3.4 Hz, 1H), 3.59 (dt, J = 11.0, 4.8 Hz, 1H), 1.95–1.89 (m, 1H), 1.81–1.76 (m, 1H), 1.75–1.70 (m, 1H), 1.65–1.54 (m, 3H). ^{13}C NMR (150 MHz, $CDCl_3$): δ 172.91, 138.27, 134.31, 133.40, 131.95, 131.56, 128.89, 126.26, 125.57, 124.48, 124.44, 97.84, 69.16, 62.17, 30.56, 25.47, 19.28. IR (KBr): 2934, 2870, 2630, 1689, 1593, 1571, 1511, 1458, 1286, 1256, 1124, 1066, 1039 cm^{-1}. HRMS (EI): *m*/*z* calcd for $C_{17}H_{18}O_4$ $[M]^+$: 286.1205, found: 286.1199.

3.11.4.4 Synthesis of Racemic Rotaxane 10

***Racemic*rotaxane 16** To a $CHCl_3$ solution (20 ml) of ring component **20** (7.2 g, 10.3 mmol) was added axis component **19** (2.8 g, 7.0 mmol). The mixture was stirred at 0 °C for 45 min, which turned to a transparent solution. Endcap component **23** (2.8 g, 9.8 mmol), *N*,*N*′-dicyclohexylcarbodiimide (5.8 g, 27.9 mmol) and tributylphosphine (345 μl, 1.4 mmol) were successively added to the mixture. The reaction mixture was stirred at 0 °C for 15 h. Then water was added to the reaction mixture, and the mixture was extracted with $CHCl_3$. The combined organic extract was washed successively with water and brine, and an organic phase separated was dried over Na_2SO_4, filtered and evaporated to dryness under a reduced pressure. The residue was subjected to short column chromatography (SiO_2, MeOH/$CHCl_3$ 1/99 to 1/20), affording a mixture of rotaxane **16** and recovered ring component **20**, which was directly used for the next step.

***Racemic*rotaxane 44 (mixture of conformers)** To a MeCN solution (70 ml) of the mixture containing rotaxane **16** was added triethylamine (19.4 ml, 139 mmol) and acetic anhydride (6.6 ml, 70 mmol) at room temperature. The reaction mixture was stirred at 40 °C for 3 h. The mixture was evaporated to dryness under a reduced pressure, and the residue was diluted with AcOEt. The mixture washed successively with water and brine, and an organic phase separated was dried over Na_2SO_4, filtered and evaporated to dryness under a reduced pressure. The residue was subjected to column chromatography (SiO_2, hexane/AcOEt 1/2 to MeOH/AcOEt 1/199), affording **44** (4.8 g, 53% yield, 2 steps) as yellow amorphous, which is a mixture of diastereomers derived from a mechanically planar chirality and a chirality of THP group. ^{1}H NMR (600 MHz, $CDCl_3$): δ 9.01 (s, 1H), 8.86–8.83 (m, 1H), 8.43–8.42 (m, 1H), 8.05–8.03 (m, 2H), 7.97–7.95 (m, 2H), 7.78–7.70 (m, 5H), 7.66–7.65 (m, 1H), 7.52–7.50 (m, 1H), 7.46–7.43 (m, 1H), 7.35–7.34 (m, 1H), 7.08–7.02 (m, 2H), 6.99–6.98 (m, 1H), 6.87–6.85 (m, 1H), 6.79–6.77 (m, 3H), 6.74–6.69 (m, 3H), 6.27–6.22 (m, 2H), 4.97–4.93 (m, 1H), 4.78–4.76 (m, 1H), 4.70–4.61 (m, 2H), 4.49 (s, 1H), 4.42–4.37 (m, 2H), 4.28–4.26 (m, 2H), 4.22 (s, 1H), 4.19–4.15 (m, 1H), 4.10–4.02 (m, 4H), 3.99–3.90 (m, 2H), 3.89–3.85 (m, 1H), 3.83–3.79 (m, 1H), 3.71–3.62 (m, 4H), 3.59–3.37 (m, 7H), 3.35–3.31 (m, 1H), 3.28–3.24 (m, 1H), 2.25–2.22 (m, 6H), 2.13–2.12 (m, 3H), 1.99–1.98 (m, 3H), 1.92–1.85 (m, 1H), 1.79–1.66 (m, 2H), 1.63–1.59 (m, 1H), 1.57–1.52 (m, 2H). ^{13}C NMR (mixture of conformers) (150 MHz, $CDCl_3$): δ 170.89, 170.83, 170.81, 167.81, 149.58, 149.55, 148.28, 148.25, 148.22, 148.18, 147.99, 141.70, 141.66, 138.34, 137.88, 137.80, 137.43, 137.20, 136.98, 136.96, 136.89, 136.86, 136.39, 134.86, 134.65, 134.61, 133.92, 132.90, 132.87, 132.54, 131.84, 131.75, 131.60, 131.23, 130.06, 130.02, 129.95, 129.76, 129.61, 129.53, 129.00, 128.97, 128.92, 128.79, 128.40, 128.36, 127.71, 127.16, 126.90, 126.85, 125.89, 125.56, 125.52, 125.33, 124.85, 124.81, 124.15, 123.99, 123.93, 123.48, 122.61, 120.19, 118.00, 115.91, 115.86, 111.30, 111.11, 97.78, 97.72, 71.85, 71.55, 70.98, 70.85, 70.69, 70.25, 70.16, 69.72, 69.63, 69.56, 69.50, 69.35, 69.32, 69.21, 68.15, 68.13, 67.75, 67.66, 62.03, 61.97, 50.45, 49.97, 47.47, 47.34, 30.53, 30.51, 25.39, 24.68, 21.67, 21.15, 21.11, 19.30, 19.26. IR (neat): 3009, 2925, 2874, 1708, 1639, 1597, 1545, 1504, 1452, 1419, 1366, 1252, 1172, 1126, 1061, 1035, 755 cm^{-1}. HRMS (ESI): *m*/*z* calcd for $C_{72}H_{79}N_3O_{18}SNa$ $[M + Na]^+$: 1328.4972, found: 1328.4988.

***Racemic*rotaxane 46** To a THF solution (85 ml) of rotaxane **44** (4.6 g, 3.5 mmol) was added an aqueous solution (25 ml) of KOH (0.48 g, 8.4 mmol) at room temperature. The reaction mixture was stirred at 40 °C for 4 h. A part of solvent was evaporated under a reduced pressure, and the mixture was extracted with $CHCl_3$. Thc combined organic extract was washed successively with water and brine, and an organic phase separated was dried over Na_2SO_4, filtered and evaporated to dryness under a reduced pressure. The residue was directly used for the next step.

***Racemic*rotaxane 10 (mixture of conformers)** To a EtOH/THF (35 ml/35 ml) solution of rotaxane **46** was added pyridinium *p*-toluenesulfonate (17.6 g, 70 mmol) at 30 °C. The reaction mixture was stirred at 30 °C for 19 h. The reaction mixture was evaporated to dryness under a reduced pressure, and the residue was diluted with $CHCl_3$. The mixture was washed successively with water and brine, and an organic phase separated was dried over Na_2SO_4, filtered and evaporated to dryness under a reduced pressure. The residue was subjected to short column chromatography (SiO_2, hexane/AcOEt 1/5 to 0/1), then subjected to preparative TLC (SiO_2, MeOH/$CHCl_3$ 1/19) to allow isolation of **10** (1.53 g, 37% yield, 2 steps) as yellow amorphous. HPLC conditions: CHIRALPAK IC column, eluent EtOH/CH_2Cl_2 1/60, flow rate 0.7 ml/min, temperature 20 °C, 18.8 min (minor), 24.5 min (major). $[\alpha]_D^{23}$ −23.4 (major enantiomer, 99% ee, *c* 0.72, $CHCl_3$). ^1H NMR (600 MHz, $CDCl_3$): δ 8.83–8.80 (m, 1H), 8.70–8.66 (m, 1H), 8.03 (d, *J* = 7.9 Hz, 1H), 7.96 (d, *J* = 7.9 Hz, 1H), 7.83–7,81 (m, 2H), 7.77–7.74 (m, 1H), 7.71–7.64 (m, 2H), 7.62–7.59 (m, 1H), 7.55 (s, 1H), 7.46–7.42 (m, 4H), 7.22–7.18 (m, 2H), 7.10–7.03 (m, 2H), 6.97 (d, *J* = 8.2 Hz, 1H), 6.88–6.85 (m, 1H), 6.77–6.74 (m, 3H), 6.66–6.63 (m, 3H), 6.25–6.20 (m, 2H), 4.80–4.78 (m, 2H), 4.55–4.44 (m, 2H), 4.37–4.35 (m, 1H), 4.33–4.29 (m, 1H), 4.26–4.24 (m, 1H), 4.22–4.18 (m, 2H), 4.14–4.08 (m, 1H), 4.02–3.96 (m, 4H), 3.90–3.76 (m, 3H), 3.72–3.68 (m, 1H), 3.63–3.59 (m, 3H), 3.54–3.43 (m, 5H), 3.40–3.33 (m, 2H), 3.29–3.26 (m, 1H), 3.21–3.17 (m, 1H), 2.25–2.22 (m, 6H), 2.11–2.07 (m, 3H), 1.74 (br, 1H). ^{13}C NMR (mixture of conformers) (150 MHz, $CDCl_3$): δ 171.06, 170.96, 168.24, 148.30, 148.26, 148.17, 148.15, 148.10, 141.59, 141.58, 139.56, 139.49, 138.47, 137.98, 137.76, 137.37, 137.19, 136.37, 134.91, 133.98, 133.81, 132.79, 132.75, 132.49, 132.19, 131.70, 131.57, 131.42, 131.18, 131.14, 131.11, 131.10, 131.07, 130.06, 129.61, 129.28, 129.25, 129.04, 128.90, 128.50, 128.46, 128.33, 128.24, 127.68, 127.64, 127.12, 125.93, 125.32, 125.12, 124.98, 124.95, 124.26, 123.94, 123.89, 123.77, 123.73, 123.08, 122.47, 122.44, 121.02, 120.22, 115.87, 115.82, 111.30, 111.17, 71.79, 70.87, 70.63, 70.30, 70.21, 69.87, 69.83, 69.79, 69.73, 69.68, 69.63, 69.17, 68.07, 67.89, 67.77, 65.64, 50.52, 50.03, 47.58, 47.46, 21.66, 21.64, 21.22, 21.17. IR (neat): 3011, 2922, 2876, 1705, 1626, 1603, 1543, 1505, 1451, 1369, 1251, 1127, 753 cm^{-1}. HRMS (ESI): *m/z* calcd for $C_{65}H_{69}N_3O_{16}SNa$ $[M + Na]^+$: 1202.4291, found: 1202.4322.

3.11.4.5 Synthesis of Ring Component 22

Ring component 22 To a DMF solution (0.66 ml) of ring component **20** (132 mg, 0.189 mmol) was added NaH (9.5 mg, 0.395 mmol), and the mixture was stirred at room temperature for 15 min. To the resulting mixture was added methyl iodide (25 µl, 0.395 mmol), and the reaction mixture was stirred at room temperature for 1 h. Then water was added to the reaction mixture, and the mixture was extracted with AcOEt. The combined organic extract was washed with water, and an organic phase separated was dried over Na_2SO_4, filtered and evaporated to dryness under a reduced pressure. The residue was subjected to column chromatography (SiO_2, MeOH/AcOEt/Hexane 1/25/25 to MeOH/$CHCl_3$ 1/50) to allow isolation of **22** (130 mg, 96% yield) as yellow oil. ^{1}H NMR (400 MHz, $CDCl_3$): δ 8.09 (d, J = 9.2 Hz, 1H), 7.66–7.58 (m, 3H), 7.48–7.34 (m, 3H), 7.27–7.24 (m, 2H), 6.89–6.86 (m, 4H), 4.35–4.33 (m, 2H), 4.31–4.29 (m, 2H), 4.16–4.14 (m, 4H), 3.96–3.90 (m, 8H), 3.85–3.79 (m, 8H), 3.45 (s, 3H). ^{13}C NMR (100 MHz, $CDCl_3$): δ 148.83, 148.73, 148.18, 148.04, 142.22, 136.04, 133.60, 131.55, 130.95, 130.91, 129.39, 128.32, 126.08, 124.99, 123.99, 123.59, 123.14, 121.42, 121.31, 117.19, 114.32, 113.98, 72.32, 70.96, 70.54, 70.48, 69.87, 69.80, 69.72, 69.41, 69.21, 68.98, 39.40.IR (neat): 2925, 2874, 1597, 1545, 1502, 1454, 1353, 1256, 1123, 752 cm^{-1}. HRMS (FAB): *m/z* calcd for $C_{35}H_{40}N_2O_{12}S$ $[M]^+$: 712.2302, found: 712.2302.

3.11.4.6 Synthesis of Racemic Rotaxane 15

***Racemic*rotaxane 18 (mixture of conformers)** To a $CHCl_3$ solution (5.2 ml) of ring component **22** (120 mg, 0.168 mmol) was added axis component **19** (94.3 mg, 0.235 mmol). The mixture was stirred at 0 °C for 5 h. To the mixture was successively added endcap component **23** (67.3 mg, 0.235 mmol), tributylphosphine (83 µl, 0.336 mmol) and *N,N*'-dicyclohexylcarbodiimide (139 mg, 0.672 mmol). The reaction mixture was stirred at 0 °C for 14 h. The reaction mixture was filtered, and evaporated to dryness under a reduced pressure. The residue was subjected to a preparative TLC (SiO_2, MeOH/$CHCl_3$ 1/19) in order to separate rotaxane **18** from ring component **22**. The resulting product containing **18** was directly used for the next step.

***Racemic*rotaxane 48 (mixture of conformers)** To a MeCN solution (940 µl) of **18** was added triethylamine (940 µl, 6.72 mmol) and acetic anhydride (320 µl, 3.36 mmol) at room temperature. The reaction mixture was stirred at 40 °C for 1 h. The mixture was evaporated to dryness under a reduced pressure. The residue was subjected to preparative TLC (SiO_2, MeOH/$CHCl_3$ 1/19), then subjected to preparative TLC (SiO_2, MeOH/AcOEt/Hexane 1/50/50), affording **48** (67.6 mg, 34% yield, 2 steps) as pale yellow amorphous, which is a mixture of diastereomers derived from a mechanically planar chirality and a chirality of THP group. ^{1}H NMR (400 MHz, $CDCl_3$): δ 9.01–8.99 (m, 1H), 8.88–8.85 (m, 1H), 8.02 (d, J = 8.3 Hz, 1H), 7.95 (d, J = 7.8 Hz, 1H), 7.88–7.85 (m, 1H), 7.76–7.71 (m, 2H), 7.63–7.56 (m, 3H), 7.53–7.49 (m, 2H), 7.44–7.42 (m, 1H), 7.39–7.35 (m, 1H), 7.30–7.27 (m, 1H), 7.19–7.17 (m, 1H), 7.11–7.02 (m, 2H), 6.97–6.95 (m, 1H), 6.88–6.85 (m, 1H), 6.79–6.76 (m, 2H), 6.73–6.68 (m, 3H), 6.66 (s, 1H), 6.23–6.22 (m, 2H), 4.96–4.91 (m, 1H), 4.78–4.75 (m, 1H), 4.69–4.64 (m, 1H), 4.59–4.45 (m, 2H), 4.40–4.33 (m, 2H), 4.25–4.13 (m,

4H), 4.08–3.77 (m, 8H), 3.71–3.22 (m, 17H), 2.25–2.22 (m, 6H), 2.12–2.09 (m, 3H), 1.92–1.83 (m, 1H), 1.78–1.51 (m, 5H). ^{13}C NMR (mixture of conformers) (100 MHz, $CDCl_3$): δ 170.99, 170.89, 167.86, 148.90, 148.27, 141.61, 138.44, 137.96, 137.78, 137.40, 137.21, 137.07, 137.00, 136.41, 135.87, 135.82, 134.90, 133.96, 133.56, 132.99, 132.95, 131.95, 131.83, 131.64, 131.34, 131.08, 130.99, 130.11, 129.66, 129.18, 129.14, 129.02, 128.88, 128.44, 127.67, 127.62, 127,16, 126.31, 125.94, 125.66, 125.60, 125.35, 125.11, 125.08, 124.88, 124.40, 124.34, 124.26, 123.96, 123.63, 122.93, 120.26, 115.73, 115.67, 111.34, 111.19, 97.88, 97.81, 71.79, 70.89, 70.70, 70.33, 70.23, 69.78, 69.72, 69.43, 69.40, 69.26, 68.20, 67.82, 67.76, 62.12, 62.05, 50.47, 50.02, 47.52, 47.36, 39.43, 30.59, 25.45, 21.68, 21.21, 21.17, 19.36, 19.32. IR (neat): 3008, 2924, 2875, 1704, 1641, 1598, 1547, 1504, 1452, 1367, 1252, 1063, 755 cm^{-1}. HRMS (ESI): *m/z* calcd for $C_{71}H_{79}N_3O_{17}SNa$ $[M + Na]^+$: 1300.5022, found: 1300.5046.

Rotaxane 15 (mixture of conformers) By a procedure similar to that for *racemic* rotaxane **10**, preparation of *racemic* rotaxane **15** was performed. Pale yellow amorphous (42 mg, 96% yield). HPLC conditions: After the kinetic resolution of *racemic* **15**, the recovered rotaxane **15** was purified by HPLC and converted to **15-OAc** by *N,N*-dimethyl-4-aminopyridine catalyst for ease of chiral HPLC analysis. $[\alpha]_D^{20}$ − 20.12 (major enantiomer, 68% ee, *c* 0.43, $CHCl_3$). ^{1}H NMR (400 MHz, $CDCl_3$): δ 8.85–8.82 (m, 1H), 8.72–8.67 (m, 1H), 8.04 (d, $J = 7.8$ Hz, 1H), 7.97 (d, $J = 8.2$ Hz, 1H), 7.89–7.86 (m, 1H), 7.74–7.67 (m, 2H), 7.63–7.57 (m, 3H), 7.52–7.36 (m, 4H), 7.26–7.18 (m, 2H), 7.12–7.04 (m, 2H), 7.00–6.98 (m, 1H), 6.88–6.85 (m, 1H), 6.77–6.74 (m, 3H), 6.67–6.63 (m, 3H), 6.27–6.24 (m, 2H), 4.81–4.79 (m, 2H), 4.61–4.53 (m, 1H), 4.50–4.30 (m, 3H), 4.27–4.18 (m, 3H), 4.15–4.07 (m, 1H), 4.04–3.94 (m, 4H), 3.91–3.77 (m, 3H), 3.73–3.60 (m, 4H), 3.57–3.33 (m, 10H), 3.30–3.26 (m, 1H), 3.22–3.17 (m, 1H), 2.25–2.21 (m, 6H), 2.12–2.10 (m, 3H). ^{13}C NMR (mixture of conformers) (100 MHz, $CDCl_3$): δ 171.07, 170.93, 168.37, 168.34, 148.87, 148.30, 148.19, 141.62, 139.61, 139.53, 138.49, 138.01, 137.78, 137.37, 137.19, 136.39, 135.92, 135.86, 134.98, 134.05, 133.58, 132.84, 132.79, 131.68, 131.50, 131.16, 131.01, 130.07, 129.63, 129.21, 129.18, 129.07, 128.93, 128.57, 128.53, 128.47, 128.44, 128.39, 128.29, 127.17, 126.34, 125.96, 125.40, 125.18, 125.12, 125.01, 124.97, 124.43, 124.37, 124.31, 123.98, 123.76, 123.72, 123.68, 122.95, 120.26, 115.76, 115.70, 111.33, 111.19, 71.85, 70.91, 70.66, 70.33, 70.24, 69.90, 69.86, 69.70, 69.22, 68.09, 67.96, 67.81, 65.71, 50.52, 50.09, 47.59, 47.43, 39.48, 21.70, 21.24, 21.18. IR (neat): 3418, 3012, 2925, 2877, 1704, 1631, 1600, 1547, 1504, 1453, 1367, 1252, 1128, 1061, 755 cm^{-1}. HRMS (ESI): *m/z* calcd for $C_{66}H_{71}N_3O_{16}SNa$ $[M + Na]^+$: 1216.4447, found: 1216.4479.

3.11.4.7 Synthesis of Ring Component 21

Compound 31 To a DMF solution (30 ml) of **28** (4.1 g, 6.9 mmol) was added CuCN (12.4 g, 138 mmol) at room temperature. The reaction mixture was stirred at 120 °C for 70 h. Water was added to the reaction mixture, and the mixture was extracted with AcOEt. The combined organic extract was washed successively with saturated aqueous solution of $NaHCO_3$, and an organic phase separated was dried over Na_2SO_4, filtered and evaporated to dryness under a reduced pressure. The residue was subjected to short column chromatography (SiO_2, Hexane/AcOEt 1/1 to MeOH/$CHCl_3$ 1/9), affording crude product containing **31** as brown oil, which was directly used for the next step.

Compound 32 To a THF solution (12.5 ml) of $LiAlH_4$ (410 mg, 10.8 mmol) was added a THF solution (12.5 ml) of **31** (1.88 g) at 0 °C. The reaction mixture was stirred at room temperature for 10 h. To the reaction mixture was added water, followed by 15% aqueous solution of NaOH and water. The resulting precipitate was filtered through celite, and a part of solvent was evaporated under a reduced pressure. The mixture was extracted with $CHCl_3$. The combined organic extract was washed with brine, and an organic phase separated was dried over Na_2SO_4, filtered and evaporated to dryness under a reduced pressure. The residue was directly used for the next step.

Ring component 21 To a CH_2Cl_2 solution (10 ml) of a mixture of **32** and triethylamine (550 μl, 3.95 mmol) was added 2-nitrobenzenesulfonyl chloride (875 mg, 3.95 mmol) at room temperature. The reaction mixture was stirred at room temperature for 1 h. The reaction mixture was diluted with $CHCl_3$. The mixture was washed successively with 1 M aqueous HCl solution and brine, and an organic phase separated was dried over Na_2SO_4, filtered and evaporated to dryness under a reduced pressure. The residue was subjected to column chromatography (SiO_2, MeOH/$CHCl_3$ 1/99) to allow isolation of **21** (2.05 g, 42% yield, 3 steps) as reddish brown amorphous. ^{1}H NMR (400 MHz, $CDCl_3$): δ 7.95 (d, J = 8.7 Hz, 1H), 7.89 (dd, J = 7.3, 0.9 Hz, 1H), 7.74 (d, J = 7.8 Hz, 1H), 7.56–7.51 (m, 2H), 7.44–7.39 (m, 2H), 7.23–7.21 (m, 2H), 6.89–6.86 (m, 4H), 5.84 (brt, J = 6.0 Hz, 1H), 4.44 (d, J = 6.0 Hz, 2H), 4.30–4.27 (m, 4H), 4.17–4.13 (m, 4H), 3.95–3.89 (m, 8H), 3.86–3.78 (m, 8H). ^{13}C NMR (100 MHz, $CDCl_3$): δ 148.82, 148.72, 147.64, 147.44, 142.09, 133.75, 133.13, 132.34, 131.16, 130.76, 129.08, 128.53, 126.65, 125.47, 124.92, 123.77, 122.23, 121.43, 121.32, 116.65, 114.29, 113.98, 72.25, 70.96, 70.52, 70.44, 69.92, 69.77, 69.71, 69.32, 69.21, 68.97, 47.85. IR (neat): 3012, 2927, 2875, 1600, 1541, 1502, 1345, 1255, 1125, 751 cm^{-1}. HRMS (FAB): m/z calcd for $C_{35}H_{40}N_2O_{12}S$ $[M]^+$: 712.2302, found: 712.2304.

3.11.4.8 Synthesis of Racemic Rotaxane 14

***Racemic*rotaxane 17** By a procedure similar to that for *racemic* rotaxane **16**, preparation of *racemic* rotaxane **17** was performed.

***Racemic*rotaxane 45 (mixture of conformers)** By a procedure similar to that for *racemic* rotaxane **44**, acylative neutralization of **17** was performed. Pale yellow oil (26 mg, 25% yield, 2 steps). ^{1}H NMR (400 MHz, $CDCl_3$): δ 9.01 (s, 1H), 8.89–8.87 (m, 1H), 8.44–8.40 (m, 1H), 8.03 (d, J = 7.8 Hz, 1H), 7.97–7.93 (m, 2H), 7.79–7.70 (m, 6H), 7.60–7.58 (m, 1H), 7.52–7.49 (m, 1H), 7.39–7.36 (m, 1H), 7.30–7.27 (m, 1H), 7.11–7.04 (m, 2H), 6.98–6.96 (m, 1H), 6.86–6.84 (m, 1H), 6.80–6.72 (m, 5H), 6.65 (s, 1H), 6.25–6.23 (m, 2H), 5.16 (s, 2H), 4.95–4.92 (m, 1H), 4.77–4.75

(m, 1H), 4.69–4.65 (m, 1H), 4.58–4.47 (m, 2H), 4.41–4.36 (m, 2H), 4.25–4.16 (m, 4H), 4.10–4.03 (m, 4H), 3.99–3.78 (m, 4H), 3.72–3.23 (m, 14H), 2.23–2.22 (m, 6H), 2.15–2.10 (m, 6H), 1.94–1.83 (m, 1H), 1.79–1.51 (m, 4H). ^{13}C NMR (mixture of conformers) (100 MHz, $CDCl_3$): δ 171.05, 170.99, 170.96, 167.88, 148.57, 148.50, 148.37, 148.35, 148.31, 148.28, 147.86, 141.69, 138.46, 137.99, 137.84, 137.46, 137.27, 137.08, 137.01, 136.46, 134.92, 134.60, 134.57, 133.99, 132.97, 132.02, 131.93, 131.50, 131.46, 131.39, 130.07, 129.65, 129.30, 129.27, 129.02, 128.90, 128.79, 128.49, 127.62, 127.25, 126.00, 125.66, 125.63, 125.43, 125.32, 124.93, 124.63, 124.56, 124.33, 123.99, 122.58, 120.30, 115.47, 115.41, 111.39, 111.24, 97.90, 97.82, 71.84, 70.90, 70.79, 70.29, 70.20, 69.82, 69.77, 69.48, 69.46, 69.24, 68.27, 67.86, 67.78, 62.15, 62.08, 51.25, 50.52, 50.00, 47.55, 47.41, 30.62, 30.61, 29.68, 25.49, 24.30, 21.73, 21.23, 21.20, 19.39, 19.34. IR (neat): 3009, 2925, 2875, 1707, 1642, 1543, 1504, 1452, 1366, 1251, 1172, 1126, 1061, 1034, 753 cm^{-1}. HRMS (ESI): *m*/*z* calcd for $C_{73}H_{81}N_3O_{18}SNa$ [M + Na]$^+$: 1342.5128, found: 1342.5146.

***Racemic*rotaxane 47** By a procedure similar to that for *racemic* rotaxane **46**, hydrolysis of **45** was performed.

Rotaxane 14 (mixture of conformers) By a procedure similar to that for *racemic* rotaxane **10**, preparation of *racemic* rotaxane **14** was performed. Pale yellow amorphous (8.4 mg, 36% yield, 2 steps). HPLC conditions: CHIRALPAK ID column, eluent EtOH/CH_2Cl_2 1/40, flow rate 0.5 ml/min, temperature 20 °C, 21.9 min (major), 24.1 min (minor). $[\alpha]_D^{20}$ −27.37 (major enantiomer, 63% ee, *c* 1.21, $CHCl_3$). ^{1}H NMR (400 MHz, $CDCl_3$): δ 8.85–8.84 (m, 1H), 8.77–8.73 (m, 1H), 8.02 (d, *J* = 8.3 Hz, 1H), 7.94 (d, *J* = 8.3 Hz, 1H), 7.89–7.85 (m, 1H), 7.74–7.66 (m, 4H), 7.57–7.54 (m, 1H), 7.45–7.30 (m, 4H), 7.22–7.09 (m, 3H), 7.03–6.95 (m, 2H), 6.88–6.85 (m, 1H), 6.79–6.72 (m, 3H), 6.69–6.66 (m, 3H), 6.25–6.22 (m, 2H), 6.03–5.92 (m, 1H), 4.79–4.77 (m, 2H), 4.52–4.33 (m, 6H), 4.26–4.11 (m, 4H), 4.06–3.99 (m, 4H), 3.90–3.80 (m, 3H), 3.73–3.68 (m, 1H), 3.65–3.45 (m, 8H), 3.40–3.21 (m, 4H), 2.25–2.22 (m, 6H), 2.12–2.09 (m, 3H). ^{13}C NMR (mixture of conformers) (100 MHz, $CDCl_3$): δ 171.20, 170.96, 168.17, 148.41, 148.24, 148.20, 148.18, 148.16, 147.58, 141.53, 141.44, 139.65, 139.59, 138.48, 138.00, 137.78, 137.36, 137.17, 136.34, 134.92, 134.00, 133.91, 133.09, 133.06, 132.84, 132.82, 132.34, 132.30, 131.73, 131.61, 131.36, 131.19, 130.96, 130.84, 129.99, 129.59, 129.05, 128.92, 128.88, 128.82, 128.67, 128.64, 128.56, 128.54, 128.20, 128.10, 127.13, 126.95, 126.89, 125.95, 125.62, 125.39, 124.97, 124.93, 124.29, 124.16, 124.07, 123.96, 123.71, 123.67, 122.03, 120.27, 115.14, 111.34, 111.22, 71.75, 70.86, 70.67, 70.26, 70.21, 69.91, 69.77, 69.70, 69.12, 69.10, 68.13, 67.91, 67.78, 65.65, 50.49, 50.04, 47.93, 47.89, 47.53, 47.41, 21.67, 21.23, 21.18. IR (neat): 3441, 3011, 2925, 2877, 1702, 1630, 1541, 1504, 1452, 1344, 1251, 1128, 1061, 753 cm^{-1}. HRMS (ESI): *m*/*z* calcd for $C_{66}H_{71}N_3O_{16}SNa$ [M + Na]$^+$: 1216.4447, found: 1216.4475.

3.11.4.9 Characterization Data for Rotaxanes 10-OAc, 14-OAc, 15-OAc

Rotaxane 10-OAc (mixture of conformers) Yellow amorphous. HPLC conditions: CHIRALPAK IC column, eluent $EtOH/CH_2Cl_2$ 1/60, flow rate 0.7 ml/min, temperature 20 °C, 9.2 min (major), 12.6 min (minor). $[\alpha]_D^{23}$ + 9.89 (major enantiomer, 40% ee, *c* 1.16, $CHCl_3$). ^1H NMR (600 MHz, $CDCl_3$): δ 9.02–9.00 (m, 1H), 8.88–8.86 (m, 1H), 8.02 (d, *J* = 8.2 Hz, 1H), 7.94 (d, *J* = 7.9 Hz, 1H), 7.83–7.81 (m, 2H), 7.75–7.69 (m, 3H), 7.63–7.59 (m, 1H), 7.56–7.55 (m, 1H), 7.48–7.42 (m, 3H), 7.35 (br, 1H), 7.25–7.24 (m, 1H), 7.20–7.18 (m, 1H), 7.12–7.07 (m, 1H), 7.03–7.01 (m, 1H), 6.97–6.95 (m, 1H), 6.89–6.86 (m, 1H), 6.76–6.73 (m, 3H), 6.68–6.66 (m, 3H), 6.24–6.20 (m, 2H), 5.26–5.24 (m, 2H), 4.55–4.44 (m, 2H), 4.38–4.33 (m, 2H), 4.25–4.12 (m, 4H), 4.07–4.00 (m, 4H), 3.92–3.77 (m, 3H), 3.71–3.68 (m, 1H), 3.63–3.60 (m, 3H), 3.54–3.36 (m, 6H), 3.33–3.29 (m, 1H), 3.25–3.22 (m, 1H), 2.25–2.22 (m, 6H), 2.13–2.07 (m, 6H). ^{13}C NMR (mixture of conformers) (150 MHz, $CDCl_3$): δ 171.01, 170.89, 167.77, 148.35, 148.27, 148.21, 148.16, 141.68, 138.49, 138.01, 137.78, 137.40, 137.38, 137.26, 136.46, 134.99, 134.45, 134.36, 134.05, 133.83, 133.14, 132.54, 132.23, 132.07, 131.88, 131.79, 131.24, 131.18, 131.14, 130.21, 129.77, 129.33, 129.05, 128.91, 128.67, 128.64, 127.82, 127.78, 127.17, 125.96, 125.85, 125.82, 125.54, 125.51, 125.32, 125.19, 124.75, 123.95, 123.16, 122.49, 121.13, 120.25, 115.93, 115.86, 111.30, 111.15, 71.82, 70.95, 70.69, 70.41, 70.32, 69.83, 69.77, 69.28, 68.20, 67.98, 67.85, 66.94, 50.52, 50.04, 47.58, 47.45, 21.70, 21.25, 21.20, 21.05. IR (neat): 2958, 2926, 2874, 1731, 1631, 1602, 1543, 1506, 1453, 1370, 1252, 1170, 1125, 1065, 753 cm^{-1}. HRMS (ESI): *m*/*z* calcd for $C_{67}H_{71}N_3O_{17}SNa$ $[M + Na]^+$: 1244.4396, found: 1244.4416.

Rotaxane 15-OAc (mixture of conformers) Pale yellow amorphous. HPLC conditions: CHIRALPAK ID column, eluent $EtOH/CHCl_3$ 1/540, flow rate 0.7 ml/min, temperature 20 °C, 6.6 min (minor), 7.1 min (major). $[\alpha]_D^{20}$ + 19.07 (major enantiomer, 62% ee, *c* 1.17, $CHCl_3$). ^1H NMR (400 MHz, $CDCl_3$): δ 9.02–9.00 (m, 1H), 8.90–8.86 (m, 1H), 8.03 (d, *J* = 8.2 Hz, 1H), 7.96 (d, *J* = 8.2 Hz, 1H), 7.88–7.86 (m, 1H), 7.77–7.71 (m, 2H), 7.62–7.56 (m, 3H), 7.51–7.49 (m, 1H), 7.45–7.42 (m, 2H), 7.40–7.35 (m, 1H), 7.29–7.26 (m, 1H), 7.20–7.17 (m, 1H), 7.14–7.07 (m, 1H), 7.04 (d, *J* = 7.8 Hz, 1H), 6.98 (d, *J* = 7.8 Hz, 1H), 6.88–6.85 (m, 1H), 6.76–6.73 (m, 3H), 6.70–6.67 (m, 3H), 6.25–6.22 (m, 2H), 5.26–5.24 (m, 2H), 4.60–4.52 (m, 1H), 4.49–4.46 (m, 1H), 4.40–4.34 (m, 2H), 4.27–4.12 (m, 4H), 4.07–4.00 (m, 4H), 3.94–3.78 (m, 3H), 3.72–3.61 (m, 4H), 3.53–3.36 (m, 10H), 3.33–3.29 (m, 1H), 3.26–3.22 (m, 1H), 2.25–2.22 (m, 6H), 2.12–2.09 (m, 6H). ^{13}C NMR (mixture of conformers) (100 MHz, $CDCl_3$): δ 171.06, 170.92, 167.84, 148.93, 148.90, 148.31, 148.26, 148.19, 141.65, 141.63, 138.50, 138.01, 137.79, 137.40, 137.22, 136.42, 135.92, 135.88, 135.00, 134.46, 134.38, 134.06, 133.57, 133.14, 133.12, 132.11, 131.95, 131.83, 131.69, 131.26, 131.16, 131.00, 130.20, 129.76, 129.23, 129.19, 129.07, 128.93, 128.70, 128.67, 128.50, 128.47, 127.81, 127.76, 127.19, 126.35, 125.97, 125.84, 125.81, 125.58, 125.54, 125.39, 125.17, 125.11, 124.77, 124.45, 124.39, 123.99, 123.67, 122.96, 120.27, 115.74, 115.68, 111.31, 111.16, 71.85, 70.96, 70.70, 70.41, 70.31, 69.84, 69.75, 69.29, 68.21, 68.00, 67.85, 66.95, 50.51, 50.08, 47.59, 47.42, 39.48, 21.71, 21.24, 21.19, 21.04. IR (neat): 2925, 2855, 1738,

1706, 1641, 1599, 1547, 1504, 1454, 1369, 1254, 1124, 757 cm^{-1}. HRMS (ESI): *m/z* calcd for $C_{68}H_{73}N_3O_{17}SNa$ $[M + Na]^+$: 1258.4553, found: 1258.4569.

Rotaxane 14-OAc (mixture of conformers) Pale yellow amorphous. HPLC conditions: CHIRALPAK ID column, eluent $EtOH/CH_2Cl_2$ 1/80, flow rate 1.0 ml/min, temperature 20 °C, 9.2 min (minor), 9.8 min (major). $[\alpha]_D^{20}$ + 19.13 (major enantiomer, 64% ee, *c* 0.79, $CHCl_3$). ^1H NMR (600 MHz, $CDCl_3$): δ 9.02 (s, 1H), 8.93–8.90 (m, 1H), 8.01 (d, $J = 8.3$ Hz, 1H), 7.92 (d, $J = 8.3$ Hz, 1H), 7.87–7.82 (m, 1H), 7.78–7.65 (m, 4H), 7.57–7.55 (m, 1H), 7.47–7.21 (m, 5H), 7.17–7.13 (m, 2H), 7.01–6.95 (m, 2H), 6.88–6.85 (m, 1H), 6.77–6.66 (m, 6H), 6.27–6.18, (m, 2H), 6.01–5.88 (m, 1H), 5.26–5.20 (m, 2H), 4.50–4.32 (m, 6H), 4.28–4.00 (m, 8H), 3.92–3.80 (m, 3H), 3.73–3.68 (m, 1H), 3.64–3.26 (m, 12H), 2.25–2.22 (m, 6H), 2.12–2.08 (m, 6H). ^{13}C NMR (mixture of conformers) (150 MHz, $CDCl_3$): δ 171.11, 170.93, 167.74, 167.73, 148.49, 148.37, 148.32, 148.29, 148.26, 147.66, 141.54, 141.51, 138.50, 138.02, 137.82, 137.40, 137.25, 136.44, 135.00, 134.55, 134.48, 134.08, 134.02, 133.18, 133.09, 133.05, 132.33, 132.28, 132.18, 131.96, 131.87, 131.32, 131.00, 130.96, 130.89, 130.14, 129.75, 129.07, 128.93, 128.77, 128.71, 128.69, 128.58, 127.85, 127.79, 127.18, 127.01, 126.94, 126.00, 125.86, 125.80, 125.64, 125.59, 125.41, 124.99, 124.94, 124.80, 124.17, 124.08, 124.00, 121.11, 120.33, 120.30, 120.28, 115.21, 111.37, 111.24, 71.79, 70.95, 70.75, 70.38, 70.34, 69.97, 69.83, 69.79, 69.74, 69.68, 69.24, 69.20, 68.26, 68.01, 67.88, 66.94, 50.54, 50.09, 48.02, 47.97, 47.56, 47.46, 21.69, 21.25, 21.21, 21.05. IR (neat): 3012, 2924, 2878, 1738, 1707, 1633, 1603, 1541, 1452, 1346, 1251, 1128, 1062, 754 cm^{-1}. HRMS (ESI): *m/z* calcd for $C_{68}H_{73}N_3O_{17}SNa$ $[M + Na]^+$: 1258.4553, found: 1258.4613.

3.11.5 NMR Spectroscopy

3.11.5.1 NMR Spectra of 10

See Fig. 3.33.

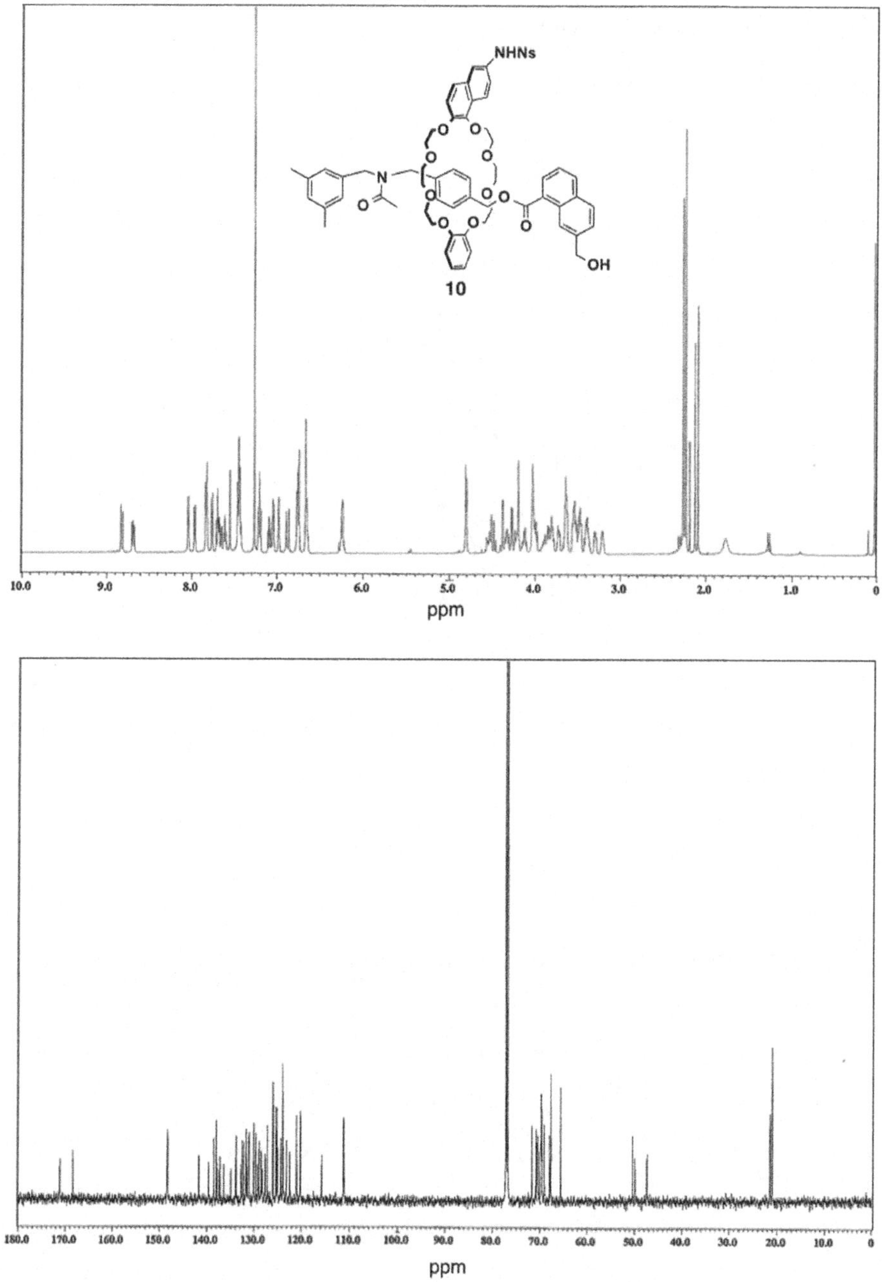

Fig. 3.33 ^{1}H NMR (600 MHz) and ^{13}C NMR (150 MHz) spectra of **10** in $CDCl_3$

3.11.5.2 NMR Spectra of 10-OAc

See Figs. 3.34, 3.35, 3.36 and 3.37.

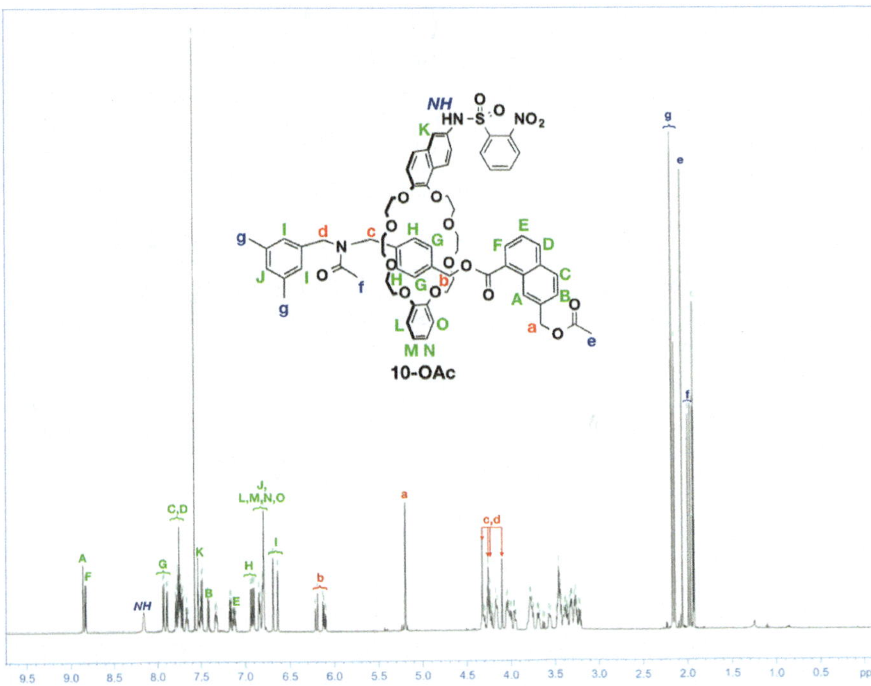

^{1}H NMR (600 MHz, CD_3CN, 20 °C): δ 8.86 (s, 1H, A), 8.84–8.83 (m, 1H, F), 8.16 (brs, 1H, *NH*), 7.94–7.89 (m, 2H, G), 7.80–7.71 (m, 5H, C+D+3H), 7.68–7.65 (m, 1H), 7.54–7.53 (m, 1H, K), 7.52–7.48 (m, 2H), 7.42–7.40 (m, 1H, B), 7.34–7.32 (m, 1H), 7.18–7.15 (m, 1H), 7.14–7.12 (m, 1H, E), 6.94–6.90 (m, 2H, H), 6.85–6.79 (m, 5H, J+L+M+N+O), 6.69–6.64 (m, 2H, I), 6.21–6.19 (m, 1H, b), 6.13–6.09 (m, 1H, b), 5.20 (s, 2H, a), 4.34–4.10 (m, 8H, c+d+4H), 4.06–3.94 (m, 4H), 3.82–3.75 (m, 3H), 3.71–3.68 (m, 1H), 3.57–3.55 (m, 1H), 3.48–3.20 (m, 11H), 2.17–2.14 (m, 6H, g), 2.06 (s, 3H, e), 1.99–1.96 (m, 3H, f).

Fig. 3.34 ^{1}H NMR spectrum (600 MHz) of **10-OAc** in CD_3CN at 20 °C

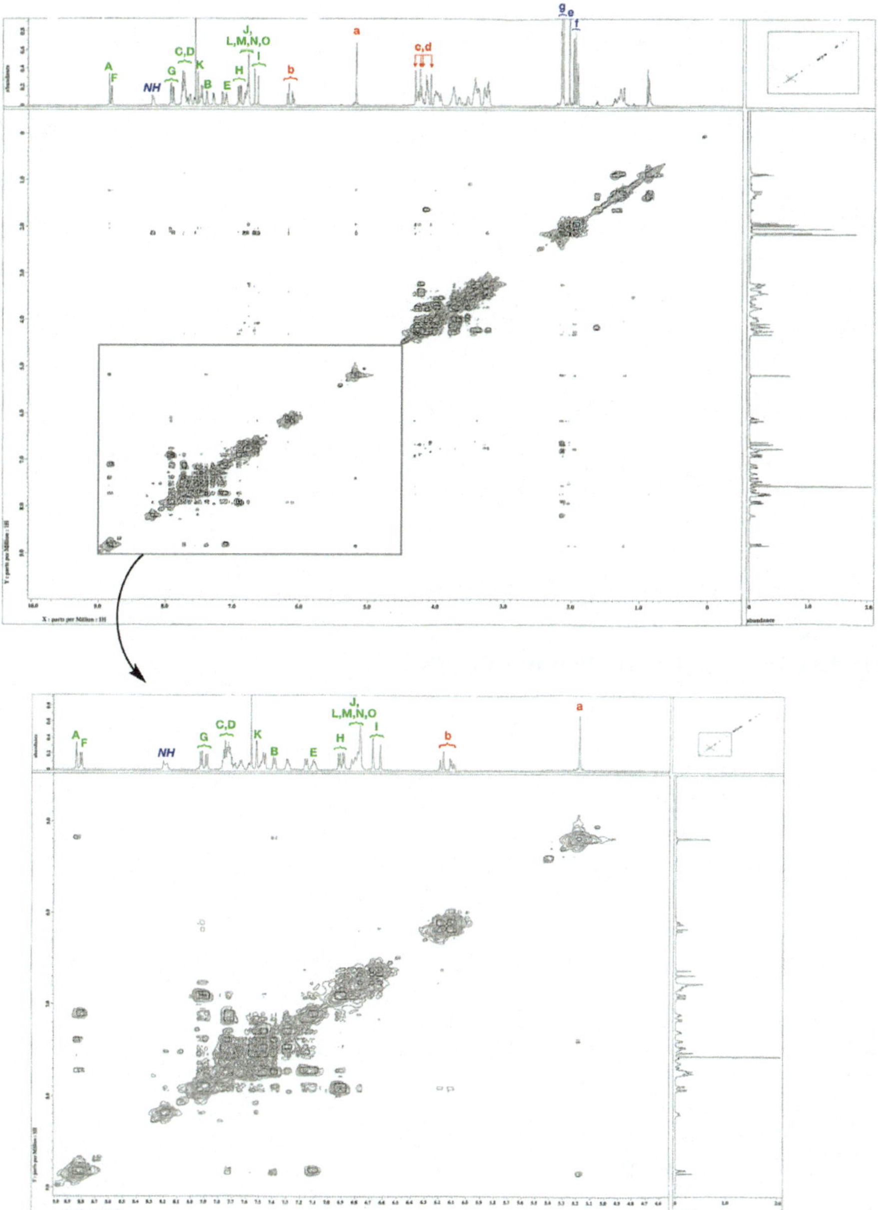

Fig. 3.35 COSY spectrum of **10-OAc** in CD_3CN

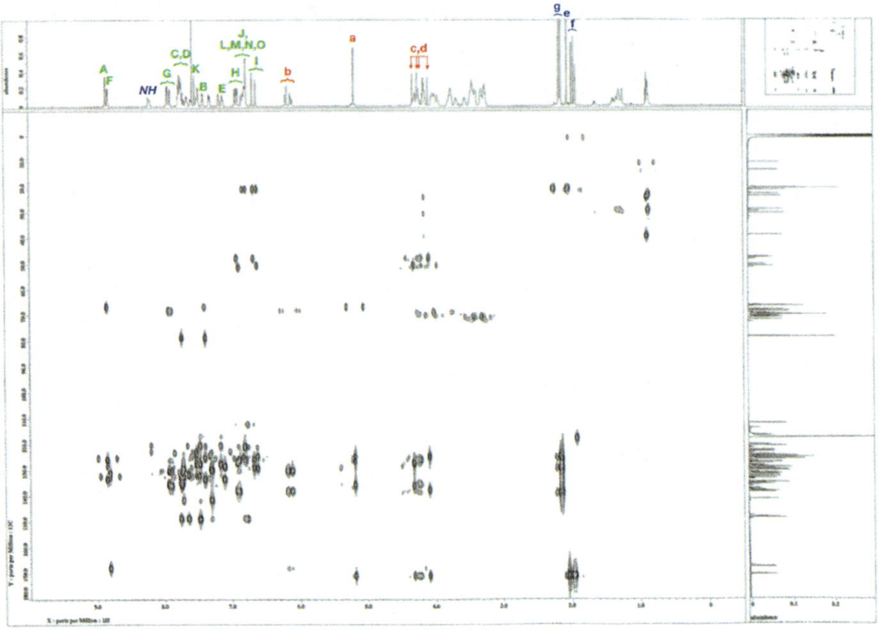

Fig. 3.36 HMBC spectrum of **10-OAc** in CD_3CN

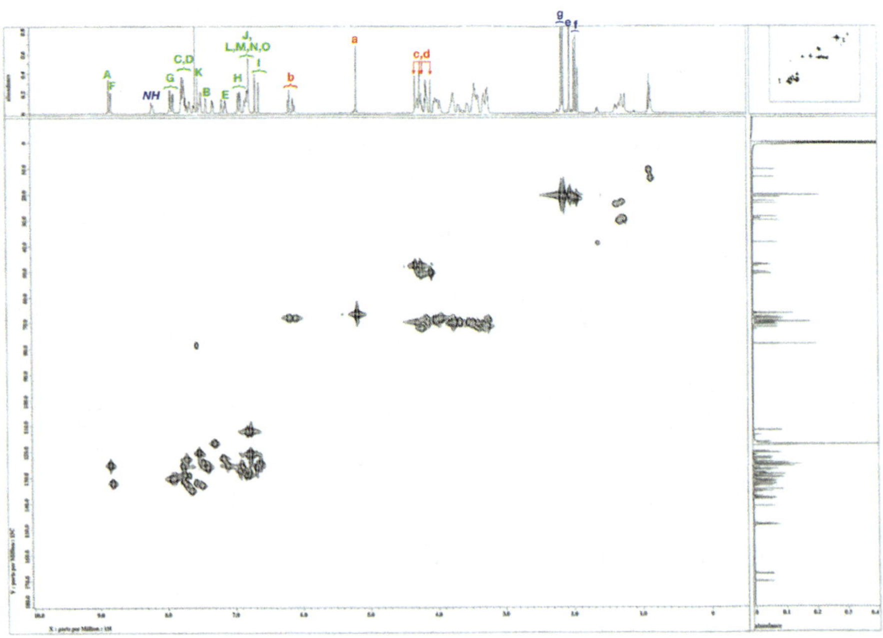

Fig. 3.37 HMQC spectrum of **10-OAc** in CD_3CN

References

1. Frisch HL, Wasserman E (1961) Chemical topology. J Am Chem Soc 83:3789–3795
2. Schill G (1971) Catenanes, rotaxanes and knots. Academic Press, New York
3. Theil A, Mauve C, Adeline MT, Marinetti A, Sauvage JP (2006) Phosphorus-containing [2]Catenanes as an example of interlocking chiral structures. Angew Chem Int Edn 45:2104–2107
4. Alvarez-Perez M, Goldup SM, Leigh DA, Slawin AMZ (2008) A chemically-driven molecular information ratchet. J Am Chem Soc 130:1836–1838
5. Tachibana Y, Kihara N, Takata T (2004) Asymmetric benzoin condensation catalyzed by chiral rotaxanes tethering a thiazolium salt moiety via the cooperation of the component: can rotaxane be an effective reaction field? J Am Chem Soc 126:3438–3439
6. Hattori G, Hori T, Miyake Y, Nishibayashi Y (2007) Design and preparation of a chiral ligand based on a pseudorotaxane skeleton: application to rhodium-catalyzed enantioselective hydrogenation of enamides. J Am Chem Soc 129:12930–12931
7. Okada Y, Miao ZH, Akiba M, Nishimura J (2006) Synthesis and characterization of chiral catenanes based on rigid Calix[4]Arene. Tetrahedron Lett 47:2699–2702
8. Mobian P, Banerji N, Bernardinelli G, Lacour J (2006) Towards the stereoselective synthesis of inherently chiral pseudorotaxanes. Org Biomol Chem 4:224–231
9. Kaida Y, Okamoto Y, Chambron J-C, Mitchell DK, Sauvage J-P (1993) The Separation of optically active copper (I) catenates. Tetrahedron Lett 34:1019–1022
10. Yamamoto C, Okamoto Y, Schmidt T, Jager R, Vogtle F (1997) Enantiomeric resolution of cycloenantiomeric rotaxane, topologically chiral catenane, and pretzel-shaped molecules: observation of pronounced circular dichroism. J Am Chem Soc 119:10547–10548
11. Schmieder R, Hubner G, Seel C, Vogtle F (1999) The first cyclodiasteromeric [3]rotaxane. Angew Chem Int Edn 38:3528–3530
12. Bordoli RJ, Goldup SM (2014) An efficient approach to mechanically planar chiral rotaxanes. J Am Chem Soc 136:4817–4820
13. Jinks MA, de Juan A, Denis M, Fletcher CJ, Galli M, Jamieson EMG, Modicom F, Zhang Z, Goldup SM (2018) Stereoselective synthesis of mechanically planar rotaxanes. Angew Chem Int Edn 57:14806–14810
14. Denis M, Lewis JEM, Modicom F, Goldup SM (2019) An auxiliary approach for the stereoselective synthesis of topologically chiral catenanes. Chem 5:1512–1520
15. Tian C, Fielden SDP, Pérez-Saavedra B, Vitorica-Yrezabal IJ, Leigh AD (2020) Single-step enantioselective synthesis of mechanically planar chiral [2]rotaxanes using a chiral leaving group strategy. J Am Chem Soc 142:9803–9808
16. Makita Y, Kihara N, Nakakoji N, Takata T, Inagaki S, Yamamoto C, Okamoto Y (2007) Catalytic asymmetric synthesis and optical resolution of planar chiral rotaxane. Chem Lett 36:162–163
17. Maynard JR-J, Goldup SM (2020) Strategies for the synthesis of enantiopure mechanically chiral molecules. Chem 6:1914–1932
18. Yoshida K, Shigeta T, Furuta T, Kawabata T (2012) Catalyst-controlled reversal of chemoselectivity in acylation of 2-aminopentane-1,5-diol derivatives. Chem Commun 48:6981–6983
19. Yoshida K, Mishiro K, Ueda Y, Shigeta T, Furuta T, Kawabata T (2012) Non-enzymatic geometry-selective acylation of tri- and tetrasubstituted α, α'-alkenediols. Adv Synth Catal 354:3291–3298
20. Yamanaka M, Yoshida U, Sato M, Shigeta T, Yoshida K, Furuta T, Kawabata T (2015) Origin of high *E*-selectivity in 4-pyrrolidinopyridine-catalyzed tetrasubstituted α, α'-alkenediols: a computational and experimental study. J Org Chem 80:3075–3082
21. Imayoshi A, Yamanaka M, Sato M, Yoshida K, Furuta T, Ueda Y, Kawabata T (2016) Insights into the molecular recognition process in organocatalytic chemoselective monoacylation of 1,5-pentanediol. Adv Synth Catal 358:1337–1344
22. Tachibana Y, Kawasaki H, Kihara N, Takata T (2006) Sequential O- and N-acylation protocol for high-yield preparation and modification of rotaxanes: synthesis, functionalization, structure, and intercomponent interaction of rotaxanes. J Org Chem 71:5093–5104

23. Keith JM, Larrow JF, Jacobsen EN (2001) Practical considerations in kinetic resolution reactions. Adv Synth Catal 343:5–26
24. Vedejs E, Jure M (2005) Efficiency in nonenzymatic kinetic resolution. Angew Chem Int Edn 44:3974–4001
25. Birman VB, Li XM (2006) Benzotetramisole: a remarkably enantioselective acyl transfer catalyst. Org Lett 8:1351–1354
26. List B (ed) (2012) Science of synthesis, asymmetric organocatalysis 1, Lewis Base and Acid Catalysts. Thieme, Stuttgart, New York
27. Kawabata T, Muramatsu W, Nishio T, Shibata T, Schedel H (2007) A catalytic one-step process for the chemo- and regioselective acylation of monosaccharides. J Am Chem Soc 129:12890–12895

Chapter 4
Appendix

Abstract Acylpyridinium salts have been known as catalytically active species in acylation reactions catalyzed by 4-pyrolidinopyridine (PPY) and its analogues. I employed PPY catalysts and acid anhydrides as acyl donors throughout the screening in this book because carboxylate anion generated from anhydride and catalyst was expected to be critically important for both selectivity and acceleration of the acylation reactions (Yamanaka et al. in J Org Chem 80:3075–3082, 2015 [1]; Kawabata et al. in J Am Chem Soc 129:12890–12895, 2007 [2]). Although relatively high catalyst loading was performed in kinetic resolution of rotaxanes (Chap. 3, Fig. 3.24), only a small amount of catalytically active species (i.e., acylpyridinium salts) was actually produced. In this chapter, I investigated the efficiency of the formation of the acylpyridinium salts depending on the acyl donor in detail (Fig. 4.1).

Keywords Formation of catalytically active species · PPY catalyst · Acid anhydride

4.1 Formation of Acylpyridinium Salts

4.1.1 Formation of Acylpyridinium Acetate Complex from PPY Catalyst and Ac_2O

First, I performed variable-temperature NMR experiment to investigate the formation of acylpyridinium acetate complex **54** generated from PPY and acetic anhydride (Ac_2O). A $CDCl_3$ solution containing acetic anhydride and PPY was monitored by NMR in the temperature range between −30 and −65 °C, and the position of equilibrium (K) was directly determined from the integration ratio of PPY to **54** at each temperature (Fig. 4.2). The thermodynamic parameters of the process were determined to be $\Delta H = -7.27$ kcal/mol and $\Delta S = -32.6$ cal/mol K. Based on these parameters, **54** were calculated to be formed in 0.12% yield in a solution of acetic anhydride (0.07 M) and PPY (0.07 M) at 20 °C (Fig. 4.1). The formation of **54** was also estimated to be only 1.0% based on the amount of PPY on the optimized conditions of kinetic resolution of rotaxane (shown in Chap. 3, Fig. 3.21, entry 5), i.e., in a solution of acetic anhydride (0.005 M) and PPY (0.015 M) at −60 °C.

A. Imayoshi, *Discrimination of Mobile Supramolecular Chirality*, Springer Theses,
https://doi.org/10.1007/978-981-16-7431-0_4

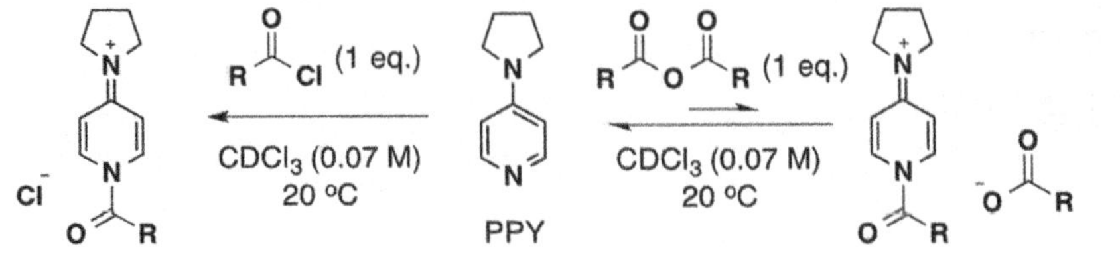

R = Me (**56**): 100%
R = Ph (**57**): 100%

R = Me (**54**): 0.12%
R = Ph (**55**): 0.0036%

Fig. 4.1 Formation of acylpyridinium salts formed from PPY and Ac_2O, Bz_2O, AcCl, and BzCl

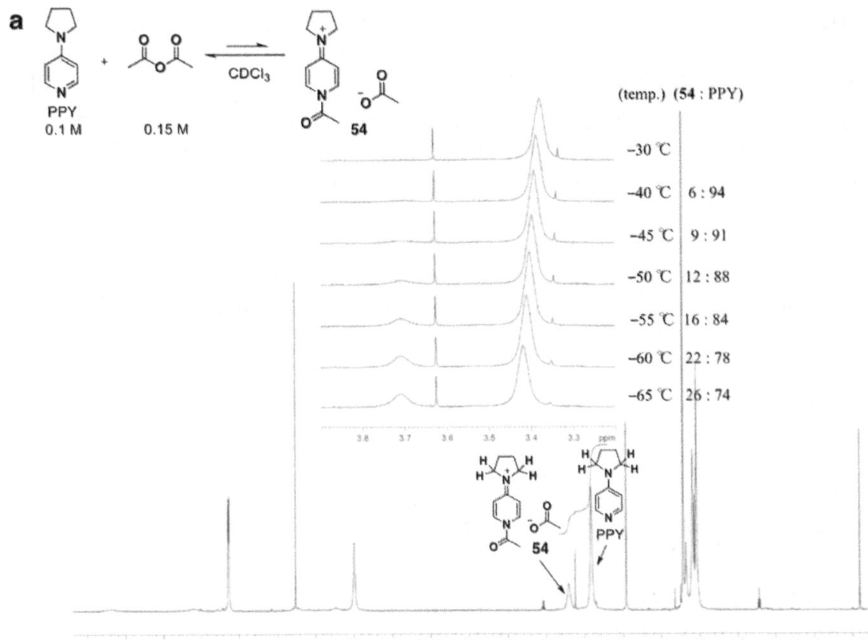

b

Temp. (°C)	Formation of **54** (%)	*K* (mol^{-1})	Δ*G* (kcal/mol)
−40	6.3	0.47	0.35
−45	8.8	0.68	0.17
−50	12	0.99	0.0053
−55	16	1.4	−0.15
−60	22	2.2	−0.33
−65	27	3.0	−0.46

ΔH = −7.27 (kcal/mol)
ΔS = −32.6 (cal/mol·K)

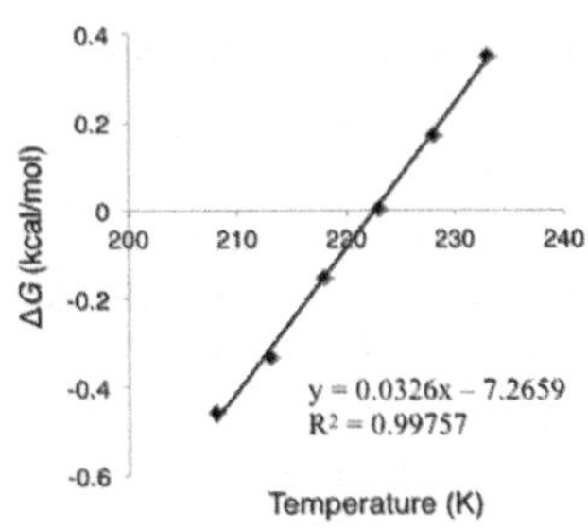

Fig. 4.2 **a** VT-NMR spectra and **b** data for the ratio of PPY to acetylpyridinium acetate complex **54**

4.1.2 *Formation of Benzoylpyridinium Benzoate Complex from PPY Catalyst and Bz_2O*

Formation of benzoylpyridinium benzoate complex **55** generated from benzoic anhydride (Bz_2O) and PPY was also investigated in a similar way to investigation of formation of acylpyridinium acetate complex **54** by variable-temperature NMR experiment in the temperature range between −45 and −70 °C (Fig. 4.3). In the case of benzoic anhydride, further less formation of acylpyridinium salts **55** was observed by the NMR spectra. The thermodynamic parameters of the process were determined to be

a

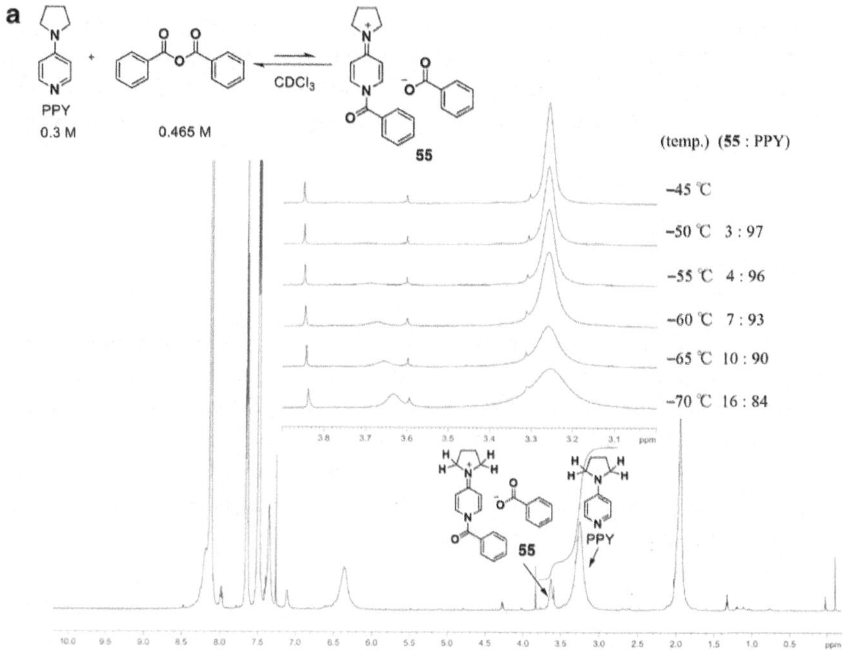

b

Temp. (°C)	Formation of **55** (%)	*K* (mol^{-1})	ΔG (kcal/mol)
−50	2.8	0.063	1.2
−55	4.3	0.10	1.0
−60	6.6	0.16	0.78
−65	9.7	0.25	0.58
−70	16	0.46	0.32

ΔH = −8.75 (kcal/mol)

ΔS = −44.8 (cal/mol·K)

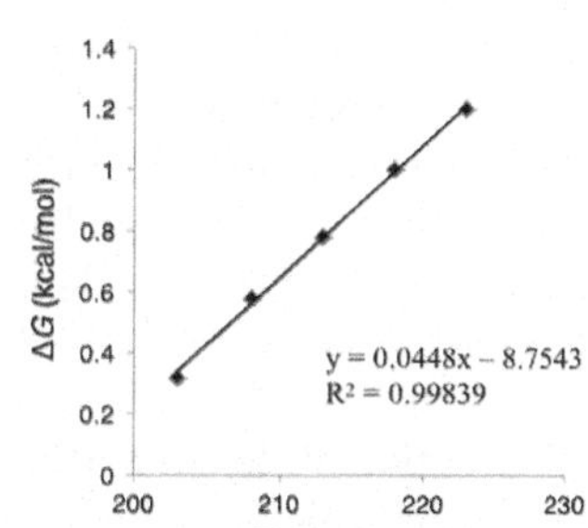

Fig. 4.3 **a** VT-NMR spectra and **b** data for the ratio of PPY to benzoylpyridinium benzoate complex **55**

$\Delta H = -8.75$ kcal/mol and $\Delta S = -44.8$ cal/mol K. Based on these parameters, **55** were calculated to be formed in 0.0036% yield in a solution of benzoic anhydride (0.07 M) and PPY (0.07 M) at 20 °C (Fig. 4.1). Only 0.08% of **55** was also found to be formed in a solution of benzoic anhydride (0.005 M) and PPY (0.015 M) at −60 °C.

4.1.3 Formation of Acylpyridinium Chloride Complexes from PPY Catalyst and AcCl or BzCl

Then, I investigated the formation of acylpyridinium chloride complexes **56** and **57** generated from PPY and the corresponding acyl chlorides by measuring ^{1}H NMR at 20 °C (Figs. 4.4 and 4.5). In the both cases of acetyl chloride (AcCl) and benzoyl chloride (BzCl), quantitative formation of acylpyridinium chloride complexes **56** and **57** were observed (Fig. 4.1).

4.2 Conclusion

The efficiency in the formation of the acylpyridinium salts changed significantly depending on the leaving group of the acylating agent. Quantitative formation of the acylpyridinium salts was observed when acyl chloride was used as an acyl donor, whereas only small amount of the acylpyridinium salts was formed when acid anhydride was used as an acyl donor. The small amount of the catalytically active species contributed greatly to the selectivity and acceleration of the acylation reactions catalyzed by PPY analogue **1**.

4.3 Experimental Section

Evaluation of the formation of acetylpyridinium acetate complex 54 from PPY and acetic anhydride by VT-NMR (Fig. 4.2) The formation of acylpyridinium acetate complex **54** from PPY and acetic anhydride was monitored by variable-temperature ^{1}H NMR. A sealed NMR tube containing 0.1 M PPY and 0.15 M acetic anhydride in $CDCl_3$ was prepared. The position of the equilibrium was directly determined from the ratio of PPY to **54** in the temperature range of −30 and −65 °C. The ratio of PPY to **54** was determined by ^{1}H NMR integration.

Evaluation of the formation of benzoylpyridinium benzoate complex 55 from PPY and benzoic anhydride by VT-NMR (Fig. 4.3) The formation of benzoylpyridinium benzoate complex **55** from PPY and benzoic anhydride was monitored by variable-temperature ^{1}H NMR. A sealed NMR tube containing 0.3 M PPY and

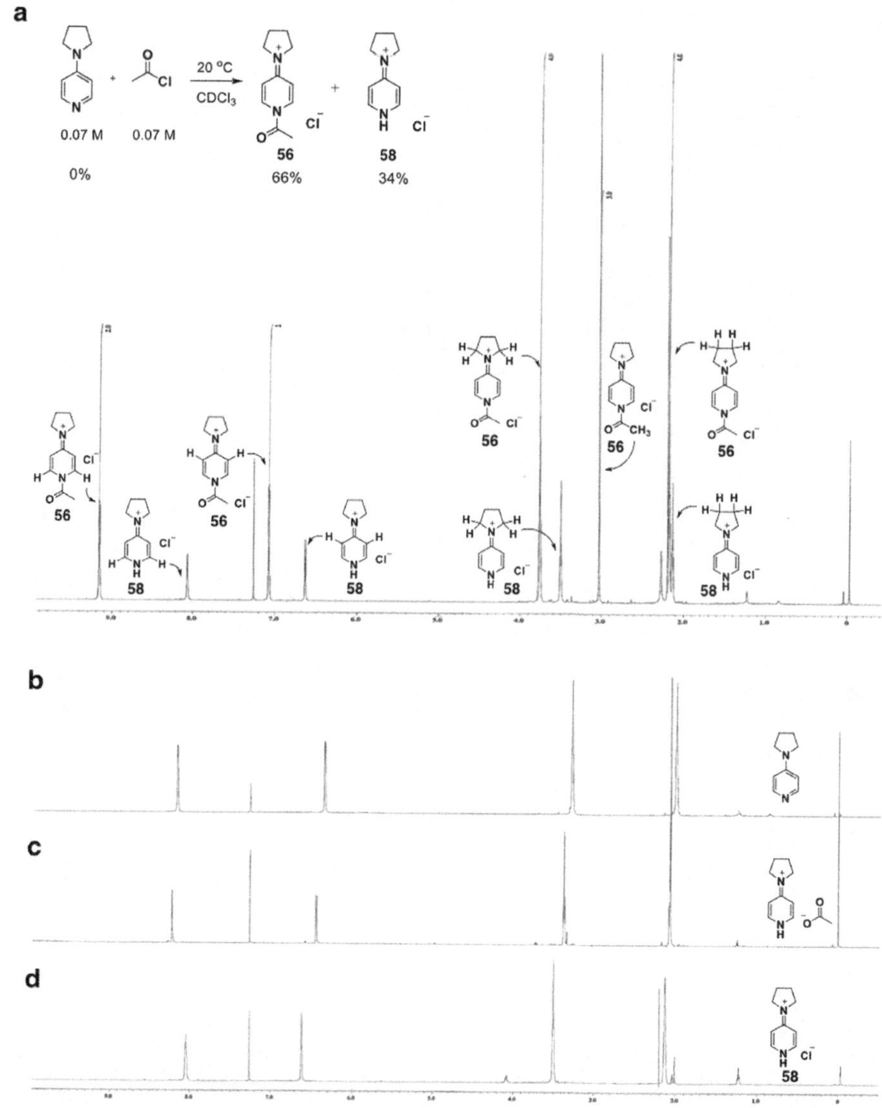

Fig. 4.4 ^{1}H-NMR spectrum for the evaluation of the formation of acetylpyridinium chloride complex **56** from PPY and acetyl chloride (AcCl), ^{1}H-NMR spectrum for **a** acetylpyridinium chloride complex **56** from PPY and acetyl chloride, **b** PPY, **c** pyridinium acetate complex, **d** pyridinium chloride complex **58**

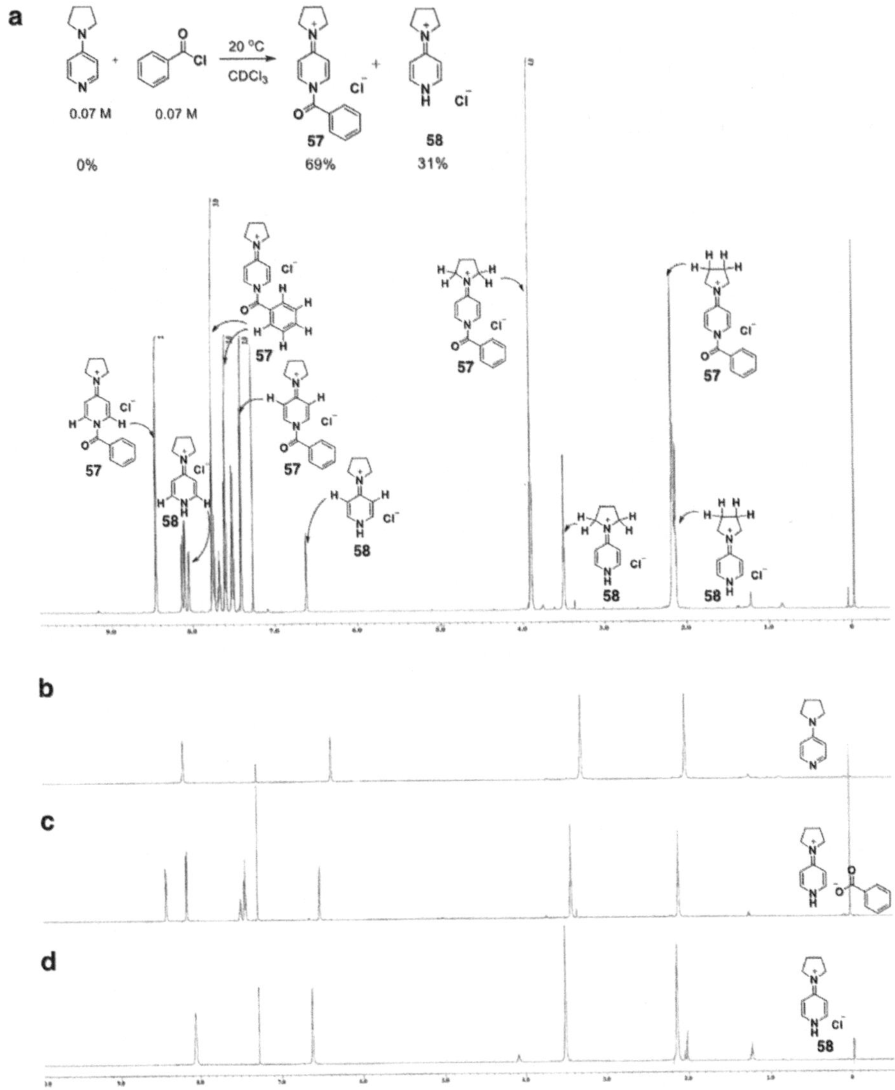

Fig. 4.5 ^{1}H-NMR spectrum for the evaluation of the formation of benzoylpyridinium chloride complex **57** from PPY and benzoyl chloride (BzCl), ^{1}H-NMR spectrum for **a** benzoylpyridinium chloride complex **57** from PPY and benzoyl chloride, **b** PPY, **c** pyridinium benzoate complex, **d** pyridinium chloride complex **58**

0.465 M benzoic anhydride in $CDCl_3$ was prepared. The position of the equilibrium was directly determined from the ratio of PPY to **55** in the temperature range of −45 and −70 °C. The ratio of PPY to **55** was determined by ^{1}H NMR integration.

Evaluation of the formation of acetylpyridinium chloride complex 56 from PPY and acetyl chloride by ^{1}H-NMR (Fig. 4.4) The formation of acylpyridinium chloride complex **56** from PPY and acetyl chloride was monitored by ^{1}H NMR. A NMR tube containing 0.07 M PPY and 0.07 M acetyl chloride in $CDCl_3$ was prepared. ^{1}H NMR spectra was obtained at 20 °C.

Evaluation of the formation of benzoylpyridinium chloride complex 57 from PPY and benzoyl chloride by ^{1}H-NMR (Fig. 4.5) The formation of acylpyridinium chloride complex **57** from PPY and benzoyl chloride was monitored by ^{1}H NMR. A NMR tube containing 0.07 M PPY and 0.07 M benzoyl chloride in $CDCl_3$ was prepared. ^{1}H NMR spectra was obtained at 20 °C.

References

1. Yamanaka M, Yoshida U, Sato M, Shigeta T, Yoshida K, Furuta T, Kawabata T (2015) Origin of high *E*-selectivity in 4-pyrrolidinopyridine-catalyzed tetrasubstituted α, α'-alkenediols: a computational and experimental study. J Org Chem 80:3075–3082
2. Kawabata T, Muramatsu W, Nishio T, Shibata T, Schedel H (2007) A catalytic one-step process for the chemo- and regioselective acylation of monosaccharides. J Am Chem Soc 129:12890–12895

Chapter 5
Conclusion and Perspectives

Abstract Production of enantiopure mechanically planar chiral rotaxanes was successfully achieved by using the chiral 4-pyrrolidinopyridine (PPY) catalysts. In our laboratory (Kawabata group), a variety of acylative molecular transformations have been developed by using the chiral PPY catalysts. These molecular transformations are based on the concept of *dynamic molecular recognition*. The chiral PPY catalysts flexibly change their recognition modes depending on the each of the substrate structures. Although the catalysts can form various transient complexes with the substrate, the reaction seems to proceed at the moment when the complex reaches the particular productive recognition mode. The dynamic and flexible recognition mode of the chiral PPY catalysts enabled discrimination of mobile and kinetic labile supramolecular chirality.

Keywords Dynamic molecular recognition · PPY catalyst · Supramolecular chirality

Discrimination of mobile supramolecular chirality was successfully achieved by using organocatalysts. The chiral PPY catalysts described in this book discriminated the dynamic chiral structures of the large mobile supramolecules in a highly selective manner by fine molecular recognition. Although the catalysts can form various transient complexes with the substrate, the reaction seems to proceed at the moment when the complex reaches the particular productive recognition mode.

The dynamic and flexible recognition mode of the chiral PPY catalysts, without depending on the steric effect, is a contrast to the traditional static and rigid recognition mode of the typical conventional catalysts. Although the conventional recognition mode is like lock-and-key model of enzyme catalysis, the dynamic and flexible recognition mode of chiral PPY catalysts is like induced-fit model of enzymatic recognition mode. The chiral PPY catalysts adapt themselves to various molecular transformations by flexibly changing their recognition modes depending on the substrate structures.

The dynamic and flexible recognition modes of the chiral PPY catalysts are somewhat similar to the mode for protein–protein interactions [1, 2]. Protein–protein interactions are crucial interactions for signalling cascades and regulation of biological events in the cell. Proteins with conformational variability have been known to form

A. Imayoshi, *Discrimination of Mobile Supramolecular Chirality*, Springer Theses,
https://doi.org/10.1007/978-981-16-7431-0_5

complexes with various different kinds of partner proteins, repeating the association/dissociation process, and finding the particular suitable interaction mode for the formation of the complex. In a similar manner to protein–protein interactions, these chiral PPY catalysts promote the reaction at the moment when the complexes reaches the most suitable recognition mode for the reaction among various possible interaction modes. The chiral PPY catalysts promote the reaction in an accelerative manner by fitting themselves to the particular active conformers of the substrates and flexibly changing their recognition modes. From such a viewpoint, the concept of dynamic molecular recognition of the chiral PPY catalysts could be applied not only to recognition of supramolecules but to controlling other giant molecules.

The concept of dynamic molecular recognition discussed in this book would provide a novel dimension in asymmetric synthesis and molecular recognition, and may contribute to further developments in a wide range of fields including material chemistry, biochemistry and medicinal chemistry beyond the field of organic chemistry.

References

1. Ozbabacan SEA, Engin HB, Gursoy A, Keskin O (2011) Transient protein-protein interactions. Protein Eng Des Sel, 1–14
2. Perkins JR, Diboun I, Dessailly BH, Lees JG, Orengo C (2010) Transient protein-protein interactions: structural, functional, and network properties. Structure 18:1233–1243

Curriculum Vitae

Ayumi Imayoshi

Current Affiliation
Graduate School for Life and Environmental Sciences,
Kyoto Prefectural University
1-5 Shimogamo Hangi-cho, Sakyo-ku, Kyoto 606-8522, Japan
Tel. +81-75-703-5603
E-mail: imayoshi@kpu.ac.jp

Former Affiliation
Institute for Chemical Research, Kyoto University
Uji, Kyoto 611-0011, Japan

Appointments

- Assistant Professor, Kyoto Prefectural University (2017–present)
- Research Fellowship for Young Scientists, Japan Society for the Promotion of Science (JSPS) (2015–2017).

Education

(Advisor: Professor Takeo Kawabata)

- Doctor of Philosophy (Ph.D.) in Pharmaceutical Sciences, Kyoto University (2016)
- Master of Science (M.S.) in Pharmaceutical Sciences, Kyoto University (2013)
- Bachelor of Science (B.S.) in Pharmaceutical Sciences, Kyoto University (2011).

A. Imayoshi, *Discrimination of Mobile Supramolecular Chirality*, Springer Theses,
https://doi.org/10.1007/978-981-16-7431-0

The manufacturer's authorised representative in the EU is Springer Nature Customer Service Centre GmbH, Europaplatz 3, 69115 Heidelberg, Germany. If you have any concerns regarding our products, please contact ProductSafety@springernature.com

Printed and bound by CPI Group (UK) Ltd, Croydon, CR0 4YY
30/09/2026
02225306-0002